# SYMBOLS & CIVILIZATION

*Science, Morals, Religion, Art*

Bell

# SYMBOLS & CIVILIZATION

*Science, Morals, Religion, Art*

by RALPH ROSS

*with ERNEST VAN DEN HAAG*

*A Harbinger Book*
HARCOURT, BRACE & WORLD, INC.
NEW YORK AND BURLINGAME

*Library of Congress Catalog Card Number: 62-21848*

*Printed in the United States of America*

c.9.66

# Acknowledgments

Since *Symbols and Civilization* is adapted from Part II of *The Fabric of Society,* written with Ernest van den Haag, it is pleasant to have the opportunity again of thanking those who were of great help with the original. Professors Robert K. Merton, of Columbia, and Abraham Kaplan, of U.C.L.A., read all of what was then Part II with, obviously, great patience and care. Because it seems mere convention to thank them for criticism and suggestion, I wish I could do more than thank them. Their criticism and suggestion were never niggling or foolish, but always serious and most carefully considered. And one cannot thank a collaborator at all. If collaboration is real, and not two or more people writing different parts of a book in utter independence, there is continuing conversation, out of which some ideas emerge, others are clarified, and still others take definitive form. Ernest van den Haag became, through these conversations—in person and by correspondence—a second intellectual conscience, and a most rigorous one.

# Contents

# Preface

For a long time I have believed that even the most elementary studies at about the college level presuppose too much, and that there should be an attempt to get basic matters as clear as possible. Before one studies science or logic or religion or the arts one should find out something about what they are and what they do, where they come from and what they are good for. Of course, all these subjects have to be studied first and only later can the elementary questions be asked. This is the case with all subjects. The elementary questions are the basic ones, and although they come first in the order of logic, they don't come first in the order of time. But that is the writer's problem, not the reader's. The reader can start, logically, from the beginning.

This book is my attempt to get clear a number of basic matters. My subject is man's ways of understanding the world and dealing with it, living with his fellows and dealing with them. From such a description, the reader might expect the following pages to be concerned with economics and politics, psychology and sociology. They are not, except incidentally. Those subjects are treated in the parent volume, *The Fabric of Society* (by Ralph Ross and Ernest van den Haag), from one part of which this book is adapted, and to which the reader is referred. Here the problems are those that underlie even economics and politics.

We are so used to talking about social influences on thought that it may seem strange to talk about thought's influence on society. Marx wrote: "It is not the consciousness of men that determines their existence, but, on the contrary, their social existence determines their consciousness." And Karl Mannheim wrote: ". . . it is not men in general who think. or even isolated individuals who do the thinking, but men in certain groups who have developed a particular style of thought in an endless series of responses to certain typical situations characterizing their common position."

Grant anything you like about the influence of society, or any of its parts, on mind, and it still neglects what seems to me more fundamental. Even the most primitive economy presupposes that a fish out of water and a speared doe will die, that if you plant potatoes you will get potatoes, not cherries, that if you eat and drink you are more likely to survive than if you don't. Underlying all human activities are the ways in which man understands his world. His experience is useless unless it is interpreted, and his interpretations range from the expectation of nightfall after a sinking sun to the doctrine of the Trinity and the atomic hypothesis.

In different cultures, much about the world is understood differently, but all cultures have interpreted the world so it is more or less intelligible. Such interpretation is at bottom communal; it requires society because it requires language and other systems of symbols. Yet the institutions and patterned activities of society depend on human communication and its underlying symbolic systems. And it is these systems that make possible the inheritance of a culture, and so its continuity.

But are Marx and Mannheim right about society and thought once language is developed and the world interpreted? Would they be right about the influences of society even on that development and interpretation? Or is George Santayana right when he says: "In the past or in the future, my language and my borrowed knowledge would have been different, but under whatever sky I had been born, since it is the same sky, I should have had the same philosophy."? There is, at the very least, one notable exception to Marx and Mannheim, and that is science. And a nice paradox is involved: if we say that science, or intellectual method of any exact sort, comes into existence under some social conditions and disappears under others, that statement may be true or false, and must itself be tested by the procedures of science.

The findings of science are trans-cultural. They are true (or false) anywhere and at any time. Since men have developed so remarkable a way of understanding and remaking the world, and since it is the overwhelmingly important phenomenon of our time, I start with science and spin out its consequences for the first eight chapters. Yet it is only an illustration of one way—today almost *the* way—of understanding the world and of making man somewhat at home in it. Hence the chapters that follow.

*Ralph Ross*

JUNE, 1962

# 1

# The Nature of Science

Science is often defined inadequately as "an organized body of knowledge." This would make cookbooks, Sears, Roebuck catalogues, and telephone books science, which they are not. Sometimes science is defined simply as rationality, but that would make much of theology and metaphysics science, which they are not. *Rationality* is logical consistency, lack of contradiction. It is to be distinguished from *reasonableness,* the quality of a mind open to arguments and evidence opposed to its beliefs: a willingness to reconsider. Rationalists can be quite unreasonable or dogmatic. Rationalist metaphysicians and theologians are often certain about premises which come from intuition or revelation. Even paranoiacs may be thought of as rationalists, for they are commonly most rigorous in reasoning. But their premises, which they cling to in spite of all evidence, are absurd.

Science is empirical, rational, general, and cumulative; and it is all four at once. Science is *empirical* in that all its conclusions are subject to test by sense experience. Observation is the base on which science rests, but scientific observation is more than keeping one's eyes open. It is observation, made by qualified observers under controlled conditions, of those things which confirm or disconfirm, verify or refute a theory. Sherlock Holmes could tell by the stains on a vest what a man had eaten for breakfast. From a number of such observations he arrived at a theory about why and how a particular crime was committed. This procedure is excellent for detection but insufficient for science, because it yields only knowledge of particular events. Science would go on to ask why and how crime, not a particular crime, is committed. Science uses facts to test general theories and general theories to make predictions about particular facts.

Scientific observation may be made of things as they exist, like the color of an apple or the temperature of the air, or it may be made of what results from an experiment. An experiment is the deliberate manipulation of conditions in order to bring about what we want to observe. If we want to test the hypothesis that a new plastic can withstand two hundred pounds of pressure without crumbling, we may have to create a situation in which such pressure is applied to a piece of the plastic, because it is unlikely that the situation already exists anywhere in the world, or, if it does, that all other factors are kept constant. In some sciences, like astronomy, we do not sufficiently control the subject matter to experiment on it—although we do control the conditions of observation—and we distinguish those sciences from others, like chemistry, in which experiment is possible, by calling the latter "experimental sciences."

Although all scientific thought *ultimately* rests on observation, there are vast portions of it which are entirely *rational:* analysis of the meanings of terms, deductions from existent theories, explorations of the logical relations among concepts and among theories. Logic is applied to science constantly because logic contains the rules of valid thinking. The application of mathematics is often thought, erroneously, to be an index of the status of any science. Of course, the more it can be applied usefully within a science, the more advanced the science. For mathematics functions both as a language in which scientific laws are stated, giving them the utmost precision, elegance, and economy, and as the basis of measurement. Many of the most significant advances in physics, astronomy, and chemistry have depended on advances in and application of mathematics. Without calculus the work of Isaac Newton would have been impossible. Yet great scientific work in other fields, performed by men like Pasteur, Darwin, and Pavlov (with whose names pasteurization, evolution, and conditioned response in psychology are associated), has used little or no mathematics. This is true also of important contributions to social science, like those of Weber, Veblen, and Mosca. Nothing is gained by the use of mathematics when a subject is not measurable or sufficiently precise.

If observation is the base of science, general laws are its crown. The body of any science is a set of general laws, logically connected, from which the occurrence of particular events is predictable. Young sciences, like nineteenth-century biology, are chiefly taxonomic; that is, they organize and classify a subject matter so that there will be enough order in it for laws to be sought. Even less-advanced sciences, like contemporary anthropology and sociology, still record a host of particular and often isolated observations so that there will be material from which to generalize. One goal of science is the creation of a unified body of knowledge which will relate all the sciences to one another; thus from laws of physics and chemistry—which deal with matters that are basic to all things, organic and inorganic—one could move logically to laws of biology, psychology, and society. Ideally, all science would be logically deducible from a single law general enough to apply to everything.

Of course, we are so far from a single set of laws for all the sciences that we can scarcely go logically from physics to chemistry. We are closer to—though still far enough from—more limited ideals of the unity of science: unity on the basis of terms or of methods. There have been relatively successful attempts to create a common body of terms for the sciences, so that propositions of one science can be meaningful in others. And, despite great differences in the procedures and apparatus of the particular sciences, it is possible to approximate a statement of the most general methods of science so that they are applicable to all science.

Science is *cumulative* in that present knowledge is based on past knowledge, even when the new supersedes the old. The dullest freshman in an introductory physics course knows more about physics than Aristotle did, not because he is brighter than Aristotle but, among other things, because of the work of Aristotle. Scientific conclusions are held tentatively; if they were regarded as certainly true, inquiry would be terminated and mistakes would be enshrined as dogma. But science is self-corrective; its mistakes are eliminated by more science. This is perhaps the basic criterion that distinguishes science from all other ways proposed for attaining truth: intuition, authority, tradition, for example. Different intuitions, authorities, and traditions may contradict each other, but each will remain firmly convinced of its own truth. These methods can be corrected only from outside themselves, especially by logical criticism and new evidence. But science is corrigible by its own practice, through continued application of its method. And the recurrent criticism that science is unreliable because it is always changing misses the point. Science is reliable *because* its conclusions change in a successive approximation of truth. It does not claim absolute truth for any of its conclusions, but only probability. And at any time we can expect the probability of a scientific law to be greater than knowledge about the same subject from any other source. Further, science does not correct itself only by recognition of mistakes. Solving one problem opens a path to others, and they in turn may lead back to a reconsideration of the earlier problem. When we learn that something previously called an instinct, for example, is actually the result of cultural conditioning, we promptly ask whether the same is true of other "instincts." And what we discover about conditioning as a result of this new inquiry may lead to greater understanding of the first "instinct."

Of course, scientific growth is not always direct; there are many blind alleys, and errors may last for years. But many errors have worked out well in the end because they led to the study of related problems or accidental discovery of some truth. Examples include the political doctrine of a state of nature,[1] which led to the isolation of specifically social and political characteristics of human behavior; the theory of humors, which pointed the way to discovery of a relationship between glandular secretions and temperament; and alchemy, which in the search for ways of transmuting other substances into gold helped found the science of chemistry. As Augustus De Morgan, a nineteenth-century English mathematician, said, "Wrong hypotheses, rightly worked, have produced more useful results than unguided observation."

Many scientists rule out of court all vague and fuzzy notions. Yet science has often advanced because of, not despite, just such notions when they were bold and imaginative, or were metaphors which called attention to a previously neglected relationship. One of the great men

[1] This is the belief that men were at first nonsocial creatures who behaved on instinct and impulse. Living together in society was thought to be the result of a rational decision which was embodied in a "social contract," stating that men would give up certain of their "natural rights" or powers to each other or to a sovereign. Some political writers knew that there probably never was a state of nature, but they continued to talk about it because it provided a useful contrast to the way men live in society, and it allowed them to distinguish the specifically social elements in human behavior from those they regarded as natural, or biological.

of modern physics, Max Planck, wrote: [2] "We must never forget that ideas devoid of a clear meaning frequently gave the strongest impulse to the further development of science." Of course, we cannot in general prefer vague ideas to exact ones; but the chief value of clarity in a statement is sometimes only the ease with which we can tell that it is wrong. Clarity and precision are great virtues, but they are not the only ones. Like sincerity and honesty in moral choice, they are not enough without intelligence. Clarity may coexist with error, triteness, and tautology. Cryptic and enigmatic statement, on the other hand, should not be dismissed without a hearing; there may be something precious buried in the words. And although a statement whose meaning is not exact should not end an inquiry, it may start one. Planck believed that with many ideas in science—like the reality of the external world, and the concept of causality—the question was not whether they were true but whether they were valuable.

Planck's distinction between the value and the truth of scientific ideas is not utterly exclusive; truth, too, is a value, and in science it is the ultimate value. But there are other scientific values, and they are not inconsistent with truth. If we distinguish between science as a product and science as a process, we can see that as a product—as a body of propositions about the world—what we value are answers, and we want them to be true. But as a process, what we value are questions and methods, with their attendant insights and new concepts. And the truth of the presuppositions on which questions are based may be irrelevant to the value of the questions themselves, or of the concepts that go with them. Here the chief values are those of stimulating and guiding further inquiry.

The cumulative character of science becomes clearer in contrast with art, which does not necessarily improve. We know more about art and artistic technique than artists of the past did, but our poets, playwrights, sculptors, and musicians are not superior to Dante, Shakespeare, Michelangelo, and Bach. The reasons are simple. Every artist, no matter what he has learned from his masters, really starts from scratch. He may not only refuse to build on past artistic achievements —which make up no single body of work, anyway—but he may have to reject much of the past for the sake of his own vision. The scientific

[2] See *Philosophy of Physics*, by Max Planck, George Allen & Unwin, Ltd., London, 1936, pp. 111 ff., for a discussion of what leads to scientific theories.

novice, however, starts his thinking with the conclusions of past science. Although he may correct some of it, he accepts most of it and goes further. And he enters a laboratory complete with instruments which his predecessors did not have, instruments making new and more exact observation and measurement possible. The artist has no such aid from instruments, for a typewriter will write no better sonnets than a quill.

Too many people believe that science is only a new body of dogma; this is harmful whether it leads people to support or to oppose science. On both sides there are excessive claims. Those who are "for" science as dogma sometimes argue that science has replaced religion. They often talk and act as though science were a new religion, and the true one. They treat scientific laws as absolute and certain. They are fascinated by the paraphernalia of science—its jargon, techniques, attitudes. This is less likely to result in science than in "scientism," an inflation of science into the only intellectual discipline and the use of science as the only measure, so that any activity that seems "scientific," whether or not it results in knowledge, is acclaimed, and any nonscientific activity, whether or not it results in beauty or wisdom, is scorned.

Those who oppose science, thinking it to be dogma, often hold it responsible for all the failures and difficulties of the modern world, although most of these are irrelevant to science or are caused by too little rather than too much scientific method. And they think that the conclusions of science are important, but that its methods are not. One of their chief mistakes is to interpret scientific law as *prescriptive*, governing the universe through a system of divine legislation, rather than as *descriptive* of the regularities of phenomena. This notion may result from confusing two basic meanings of the word "law," a scientific meaning and a legal one. When scientific laws are interpreted as if they were legal, they are thought to have the power of commands. But such laws imply a legislator, who could only be God, or Nature personified. So scientific laws are conceived as results of a cosmic jurisprudence.[3]

If scientific laws were prescriptive, nature would obey them better

[3] The opposite error consists in treating actual legislation as if it were scientific, or "natural." So laws made by government are thought of as instances of the natural order and, in consequence, never to be changed.

than men obey the laws of the state, and the possibility of chance, adventure, and novelty would disappear. The laws of nature would have to have no purpose, and attain no ends, but merely regulate events. The belief that this is so is called *mechanism,* and it was popular in the late nineteenth century. Mechanism, more fully, is a philosophy of science that interprets all phenomena as mechanical movements of matter in a world of eternal and inexorable law blindly fulfilling itself, with no place for human freedom, art, imagination, or love—which are sometimes treated as illusions. Many artists, theologians, and humanists were appalled. They tried to keep science apart from art, theology, and personal relations; they denounced the mechanization of life and blamed science for the technology which they felt overwhelmed them. George Gissing, the Victorian novelist, wrote: [4] "I hate and fear 'science' because of my conviction that for a long time to come, if not forever, it will be the remorseless enemy of mankind. I see it destroying all simplicity and gentleness of life, all beauty of the world; I see it restoring barbarism under the mask of civilization; I see it darkening men's minds and hardening their hearts; I see it bringing a time of vast conflict which will pale into insignificance 'the thousand wars of old,' and, as likely as not, will whelm all the laborious advances of mankind in blood-drenched chaos." Gissing's statement sounds like prophecy in this day of atom and hydrogen bombs, but is it science itself that brings these things or what we do with science? We might better denounce our morals or our lack of wisdom than our science.

What was perhaps most terrifying about mechanism was the implication that everything in the world could be explained in terms of mechanics, that branch of physics which deals with the actions of forces on bodies. Thus there was, and is, the belief that science "reduces" mind, consciousness, and the objects of ordinary experience to purely physical processes, or to movements of electrical charges, and that it regards all else as the illusions of the senses. Humanists in denying this have characterized science as incapable of dealing with the distinctive qualities of man and his experience, and have suggested that other methods, loosely defined in terms of sympathetic imagina-

[4] Quoted by George Lundberg in *Can Science Save Us?* Longmans, Green, 1947, pp. 79-80.

tion, insight, or common sense, be employed in the investigation of the specifically human.

Humanists have been right in their opposition, even if not helpful in their constructive proposals. By now the belief they attack is usually regarded by scientists and philosophers as fallacy, called reductionism. Reductionism consists in treating an experienced phenomenon as *nothing but* some other phenomenon which has fewer or different properties, and in asserting, ultimately, that an experienced phenomenon contains no properties beyond its physical elements. So every table would *really* be only atoms. Progressive reductionism works as follows. An art object is only mass and light waves, an act of love only chemical process; the chemical is only the physical and the physical only electrical charges; therefore, the art object or act of love is only a flow of electricity.

Ernest Nagel called this a caricature of physics for, he said,[5] ". . . the sciences seek to determine the precise *conditions* under which events come into being and continue to exist . . . the sciences do not *thereby deny* the existence of any traits found in nature . . ." If the laws stating the *conditions* for the occurrence of a social event were reducible to laws of physics, we could predict on the basis of physical knowledge alone that the event would take place, and our control of events might be correspondingly increased. We might be able at will to bring about or avert the social event. Yet this reduction of the laws of society to the laws of physics would be only a statement of the physical conditions under which an event takes place. Even now, natural science can state many of the physical, chemical, physiological, and psychological conditions under which acts of art and love take place, and the social sciences can treat their social and environmental conditions. Knowledge of these conditions allows us to predict the occurrence of such acts and perhaps to institute conditions that will bring them about. But the *conditions* are not the acts themselves nor definitions of them; to treat them as equivalent is to commit the fallacy of reductionism. Neither the social nor the natural sciences can deny without absurdity the *existence* of things whose *conditions* of existence they describe.

If science is reductionist, it cuts the ground from under its own

[5] See Ernest Nagel, "Malicious Philosophies of Science," *Partisan Review*, January-February 1943, p. 47.

feet.[6] Science rests on observation, and the world we observe contains colors, sounds, odors, and tastes. A physical hypothesis about atoms, waves, or particles refers to things that are colorless, soundless, odorless, tasteless. Yet that hypothesis is tested by observation. How paradoxical it would be to insist that the physical phenomena described by the hypothesis are the only reality, for that would imply a denial that the world we observe in testing the hypothesis—the very reason we call it true—is real!

So science may state the conditions under which art and love occur, but it is left to the humanities, and to religion, to explore the *nature* of art and love. Yet this distinction between the nature of any phenomenon and the conditions of its occurrence is misleading if it is thought to be utterly exclusive. When we know a great deal about the physical, chemical, and physiological conditions under which love occurs, and about the social and psychological conditions as well, we may be said to know something about the nature of love. Phenomena cannot be understood entirely apart from the conditions of their occurrence, but neither can they be reduced entirely to those conditions. What remains to be known about love after science has said what it can is what it feels like to love and what the significance and value of love are.

If the actual knowledge acquired by science at any time were looked at a hundred years later, much of it would turn out to be only partially true, and much actually false. Only the newer body of knowledge would then be thought of as science, and it in turn will be superseded by a still newer body. Science as method, however, despite constant refinements in procedure, has been essentially the same since about 1600. If we were forced to choose between having the conclusions of science but not its method, and scientific method but not its conclusions, there would be great advantage in choosing the latter, for the former would leave us with all the errors of our day forever fixed, while the latter, although depriving us of the scientific knowl-

[6] Some scientists are, in fact, reductionist. The American psychologist John B. Watson said that thinking was nothing but subvocal talking and that in turn was only movements of the larynx. Happily, there are fewer reductionists today than in the late nineteenth and early twentieth centuries, although those who linger on get publicity disproportionate to their importance.

edge of the moment, would allow us to reach today's knowledge and to surpass it.

Scientific method—the general methods of science—is of two kinds, analytic and empirical. The analytic is linguistic and logical; it deals with ideas and meanings. The empirical is observational; it deals with matters of fact. Analytical method alone suffices for logic and mathematics, but both analysis and observation are necessary for empirical study. General methods are distinguished from procedures or techniques, which result from the application of scientific method to a particular subject matter. One application, for example, of scientific method to the subject matter of public opinion results in the questionnaire and the interview as procedures. These depend, obviously, on special characteristics of the subject: men can think and answer questions. Questionnaires and interviews would be absurd in geological investigations; there we are limited to techniques of examining rock formations, fossils, fossil casts, and fossil molds. To be sure, it would be helpful if the rocks could talk, but they cannot. As for the special instruments of particular sciences—the familiar test tubes and Bunsen burners of chemistry laboratories, microscopes for biology, telescopes for astronomy, the great cyclotrons for physics, the galvanometers and tachistoscopes for psychology—they are not parts of general scientific method, but aids in technique and procedure.

# 2

# Language and Thought

The unit of thought is the sentence or proposition. Words convey *meanings,* but there is no information, idea, or thought unless at least two meanings are related to each other. And that relation of meanings is a proposition. The word "desk" has meaning to people who speak English, but it conveys no thought. If I utter the word, perhaps I intend you to understand "That is a desk" or "There is a desk," and the occasion of the utterance may make the intended proposition clear. Otherwise you will properly ask, "Desk? Desk? What about it?"

It is usually best to approach the subject of meaning with an analysis of words. Words are artificial signs. They are *signs* in that they stand for, refer to, call attention to something; they are *artificial* in that they are man-made and mean whatever we want them to mean. This is a simple and obvious point, but it is worth making because it is so often neglected in practice. There are four basic errors in thinking about and using words. They are: (1) the belief that there are "real" meanings, not just the meanings we have agreed to assign to words; (2) the belief that words have an innate power to affect things; (3) the belief that words are in some way like their referents, the things for which they stand; (4) the belief that words must stand for things that are tangible, or at least capable of being experienced through the senses. For purposes of exposition we shall treat these errors separately, although, as will appear, they lead into each other.

First, the belief that words have "real" meanings implies that it is right to use a particular word and no other to name or describe each referent, because that particular word "really" means the referent and no other word does. There is a story that illustrates this very well.

Adam suggested to Eve one day that it would be helpful in their conversation if they could use names to stand for all the animals they encountered. So the next morning they sat in the shade of a tree and gave a name to each kind of animal as it passed by. "Let us," said Eve, "call that a lion." Adam agreed, and seeing an equally large but striped beast said, "And let's call that a tiger." The day passed happily and fruitfully, and at twilight they rose to return to their bower. Just then a little animal they hadn't seen all day scuttled by. He was small, white, and furry, and had long ears. "We've missed him," said Adam. "What do you want to call him?" Eve answered, "Let's call him a rabbit." "Why 'rabbit'?" asked Adam. "Because," said Eve, "he *looks* like a rabbit."

Eve assumed that she knew what a rabbit looked like, although nothing had yet been called a rabbit, and she suddenly shifted from giving names arbitrarily to using a word as though it were the "real" or necessary name of a thing. There are many reasons why it is better in a particular sentence to use one word rather than another with a similar meaning (for greater simplicity, precision, and economy, for example), but there is no connection in meaning that any word has with its referent beyond the connection we have decided on.

The second error, the belief in the innate power of words, assumes a natural connection between word and referent in which the word in some way controls the referent. Many primitive peoples believe that if you know a man's real name you have power over him.[1] In the well-known fairy tale of Rumpelstiltskin, for instance, the wicked dwarf of that name stole the Queen's child and the Queen could recover the baby only when she learned the dwarf's name, which put him in her power. The ancient Jews had several names for God, one of which was His sacred or "real" name. This sacred name could not ordinarily be uttered, and could be written only if the vowels were omitted, making Jahveh into Jhvh, or Yhwh. Another instance comes from magic. Magical incantations always had to be recited in a prescribed way; if they were uttered in slightly different fashion, the words lost their power.

There are many survivals in our day of this kind of word magic, or an equivalent number magic, some of them much more complex,

[1] Analogously, they are sometimes afraid to have photographs taken because an image, or effigy, in the hands of a stranger puts their very lives in his keeping.

to be sure, but some as elementary as that in Rumpelstiltskin. Many people regard the number thirteen as bad luck, with the result that most office buildings, apartment houses, and hotels in New York City have no thirteenth floor, the numbers going directly from twelve to fourteen. When the Gregorian reform of the calendar in 1582 was inaugurated by changing the date October 5 to October 15, crowds marched in the streets, shouting, "Give us back our ten days!" People today might not be so vociferous about it, but many of them would probably feel the same way.

Such phrases as "the power of words" can be used meaningfully, for words effectively used do have great power to stir emotion, to stimulate and direct action. Words can change and have changed the course of history; but this is because what they mean—plus, sometimes, the beauty and persuasiveness of their use—has moved *men* to action. The ring of slogans echoes through history: "Liberty, Equality, Fraternity," "The Rights of Man," "of the people, by the people, and for the people." But they do not change things without human mediation.

The third error about words is that a word should be similar in some way to its referent. But the word "fire" is not hot, the word "water" is not wet, and the word "iron" is not hard. Perhaps this error results from acquaintance with rhetorical or poetic devices like onomatopoeia, in which description of sounds is dramatized by using words whose sound is like what they describe. "The clang of iron" and, more elaborately, "the buzzing of innumerable bees" duplicate the sounds to which they refer. Some words, like "buzz," may have entered the language because they sound a little like the noise to which they refer, but this is not a necessary condition for meaning, and such words are obviously very limited in number.

Usually, the feeling that goes with a word derives from its meaning, not its sound. "War," we might say, "is a grim and ugly word." But is it ugly *as a word* or because of what it signifies? "Warm" is only slightly different in sound and appearance, but it evokes pleasant emotions. How firm the word "rock" sounds when we say, "Peter was the rock on which the Church was founded," and how soft it sounds when we say, "Rock-a-bye, baby, in the tree top"!

The fourth error is unfortunately common among people who have thought a great deal about language but who have little acquaintance with philosophy, logic, or science. It is the belief that a word must stand for "something" if it is to be meaningful. Mr. Stuart Chase,

**for** example (see *The Tyranny of Words*), argued that all referents must be simple and tangible things. He assumed that discussions of trends, movements, tendencies, ideals, etc., which have no simple and tangible referents, were really only discussions about words. He was therefore free to urge that no one fear fascism, because it was only a word. But no one would ever have warned people of what the *word* "fascism" could do to them; they were warned against the complex of events to which the word refers. The only danger from the word "fascism" is that it addles our brains when we use it loosely.

To demand that a word stand for "something" is to slip into the ambiguity that technical language is designed to avoid: "something" may mean an actual thing, an entity, the tangible, the perceivable; but words can also stand intelligibly for relationships or abstractions. The word "is" obviously does not stand for any actual thing; its meanings, which are various, are all relationships. Of course, we perceive things in relation. We see that Mr. Jones *is taller than* Mr. Robinson, but this is only a way of talking about what we experience; apart from objects, we never actually see the relation intended by the phrase *is taller than*. Yet this does not make the relation less real or less observable than objects, because we cannot see objects, either, apart from relations. "But," you might say, "I can see Mr. Jones apart from his relation to Mr. Robinson." To be sure; yet you cannot see Mr. Jones without seeing any relations at all. Jones is experienced as a complex of relations: his shape alone includes the relation of his height up to his width, the relation of his head to his body, of his trunk to his legs. And in an extreme example, you cannot see a single color without seeing at least one relation and remembering at least one other. If you are shown a color in a psychological laboratory it may be the color of a surface, like a card, or it may be just a beam of colored light. Still, it must have some shape, and what you see is a colored shape, or color related to shape. In addition, to see the color properly, you must identify it, and it is best—though it is not necessary—if you know its name. To identify a color, even to identify it *as* a color, you must implicitly compare it with other colors, or with perceptions that are not colors. So you remember these things and compare them to the color before you.

The fact that what we see is always a complex of relations means that when we talk of objects or of relations we are identifying parts of a whole, parts which are never seen in isolation. Just as there is no experience of Mr. Jones without the relations that are part of him, so we never see anything that we designate by the phrase *is taller than* in the absence of objects. Yet we need the phrase, as we need names for objects, in order to think and communicate. Every sentence has a verb, because we can convey meaning only by stating a relation. And even if we translate sentences into the operational language of science, we do it in the same way for the description of an object or for the statement of its relation to another object. The sentence "Mr. Jones is six feet tall" may be restated "If Mr. Jones's height is measured with a yardstick in a straight line from top to bottom, the yardstick will be applied exactly twice." And the sentence "Mr. Jones is taller than Mr. Robinson" may be treated as a prediction based on present observation and on past experience of the heights of objects. This can be translated as follows: "If Mr. Jones and Mr. Robinson are measured with a yardstick in a straight line from top to bottom, Mr. Jones's height will be marked by a higher number than Mr. Robinson's."

There are innumerable meaningful statements which contain words standing for abstractions. The language of politics, for example, is filled with terms like "justice," "liberty," "consensus." None of these is a thing that can be seen or handled. No one situation of justice, or instance of liberty, or example of consensus fully carries the meaning of the word. A *kind* of behavior (which is what justice, etc., are) is an abstraction, but the words for it can be used meaningfully and are indispensable to all except the most rudimentary thought.

There are also entirely meaningful symbols which have no instances in *sense experience* at all. The concepts just discussed are not of this type, because although there is no possible sense experience of justice itself, one can experience individual situations which conform to the concept of justice and so can regard these situations as instances of justice. But any concept that is about concepts or meanings and not about things or events or processes has no instances in experience. For example, the words "principle" and "rule" have as their instances *statements* which contain a theory or tell us how to conduct ourselves. A rule is not even a *description* of conduct; it is a *criterion* of conduct, to which we may or may not conform. The word "rule" refers to all meanings which are criteria of this sort. And a meaning is more than just a mark on paper or a particular sound in the air; these are the material means by which meanings are communicated. A word may be written in English, in shorthand, in speedwriting, or in a foreign tongue, and it may be spoken, but the same meaning is conveyed.

Although concepts may refer to imagined things like mermaids or dragons, which can never be experienced, even these are sometimes used *as if* their referents existed (even when we know they do not) because they help us organize other concepts. So we speak of a "rigid body" in physics (though there is none) because a limit or ideal to which actual bodies approximate is useful. A school of physicists has maintained that the word "atom" stands for nothing in existence; yet these same physicists accept the atomic theory of the structure of matter. Their explanation is that although there are in fact no atoms, the concept of atom permits a kind of intellectual bookkeeping which yields accurate predictions. Thus "atom" is for them an important fiction, a concept by the use of which other concepts are organized into a system from which correct predictions follow.

Now, it is this ability to use an abstraction *in relation* to experience that marks the abstraction as meaningful. Abstractions need not signify particular things nor need they have particulars as instances, as *a* chair is an instance of "chair." If they can be used to organize or direct thought, or to predict events, abstractions are related to experience and they enter into ideas which can be tested. Even an imaginary object can be related to other imaginary objects and to real ones. We may say of a lizard with a ferocious appearance that it looks like a miniature dragon. And our statement can be tested by comparing the lizard with a standard drawing of a dragon. But there *are* abstractions that cannot be related to observable things and are "sound and fury, signifying nothing." One cannot discover such abstractions by the words that stand for them but only in the use made of them. "Happiness," "the public weal," "national harmony," for example, are words often used so vaguely that they conceal rather than reveal thought, because they are not clearly related to any identifiable experience or state of affairs. These "floating" abstractions, unanchored to experience, may be attacked as meaningless, but it is sheer confusion to go on from these to attack all abstractions.

It is difficult in ordinary language to say what we mean clearly enough to serve even everyday purposes. And it is nearly impossible to use ordinary language for the purposes of science. Words are used in so many related ways that over the centuries they develop what seem to be different and even inconsistent meanings. For example, today speakers are "heckled" at public meetings. Originally "heckle" meant an instrument for combing hemp. "Heckle" was a variant of "hackle" or "hack," to cut, or a cutting instrument, whence "hatchet." "To heckle" meant to cut roughly, to hack at. The current figurative use makes "heckle" *seem* to have two entirely different meanings.

The use of a word in different meanings is valuable to poetry but fatal to science, at least to science as product, and to logical analysis. For logic and science, every word should have a consistent single meaning, a clear and precise one, and for that they need a special use of language.

In the "Ode on a Grecian Urn," John Keats writes about the figures painted on the surface of the urn. There are young men playing music on pipes and maidens dancing or fleeing. The second stanza starts

> *Heard melodies are sweet, but those unheard*
> *Are sweeter; therefore, ye soft pipes, play on;*
> *Not to the sensual ear, but, more endear'd,*
> *Pipe to the spirit ditties of no tone:*

"Soft pipes" is ambiguous. Does "soft" mean sweet or romantic or idyllic music rather than dramatic or triumphant music? Or does "soft" mean quietly or low rather than raucously or loud? Are the melodies "unheard" because they are only imagined, or are the pipes playing so low that their sounds are below our auditory threshold; if we concentrate on the picture, can we nearly hear the music? Probably these separate but related meanings are all intended at once, and the poem gains in depth without ever becoming vague. But if Keats were making an expository statement, he would have to distinguish the meanings and state them separately.

We can often tell which of several meanings is intended by examining the sentence or paragraph in which a word is used. The word "sharp" may mean "keen-edged" when applied to the blade of a knife; it is also used metaphorically to describe an incisive intelligence, as is the word "keen." Business dealings are "sharp" when they serve the purposes of the man who engages in them but are not quite fair. And colloquially the word "sharp" may be used to mean that someone's dress is overelegant.

When science cannot wash a word clean of all meanings but one, it has to introduce another word to carry that one meaning. But this is by no means the only reason for coining new terms. New interests and ideas lead to the perception of new things for which there are as yet no words. Not only do we find actual things that no one knew about, but we have new emphases and organizations of meaning. When psychologists began to classify abnormal behavior, they used the word "agoraphobia"—composed of the Greek words *agora,* or "market place," and *phobia,* "fear"—to stand for a morbid or abnormal fear of open spaces.

One great advantage offered by technical language is the use of single words and small phrases to replace long phrases and sentences, thus providing a sort of intellectual shorthand. The word "culture" in contemporary anthropology came into common use to replace phrases like "the learned and transmitted habits of a people." As another example, interest in the general theory of signs and language has created a new vocabulary of technical terms. The whole field is called *semiotic* and is divided into three parts: (1) investigation of a language is called *pragmatics* when it describes the speaker, his relation to hearers, his intention, his social status and role; (2) investigation of a language is called *semantics* if the meanings, or things referred to, are the subject matter; (3) investigation of a language is called *syntactics* or *logical syntax* if the subject matter is the relationships among expressions in a language, rules for the construction of sentences, for example, or rules for the deduction of one sentence from another.

There are, of course, possible abuses of almost every good thing, and the chief abuses of technical language are jargon and the masking of muddle and emptiness. Scientists frequently write jargon when they say perfectly simple things, even banalities, in language so technical as to seem profound. Some people attain such facility in manipulating technical terms that they deceive themselves and others into thinking that they have made precise statements when what they are saying is confused or meaningless.

A well-known work in social psychology contains the sentence: "Since two lovers presumably have greater instigation to affectional behavior involving each other than two nonlovers have, interference with this behavior by one of the former produces a more serious frustration than would be the case with the latter." This seems to mean: "A man is more frustrated when his love, or its expression, is thwarted than he is when a weaker emotion is thwarted." Not a startling conclusion! And the more technical sentence suffers from its language because lovers may not "have greater instigation to affectional behavior" than two nonlovers when the nonlovers hate or despise each other. The sentence is therefore less likely to be true than if it had simply compared love with a weaker emotion.

Yet without the use of technical words, science might find itself in a situation common to ordinary discussion, in which verbal arguments are confused with real ones. Humpty Dumpty told Alice: "When *I* use a word, it means just what I choose it to mean—neither more nor less." And when she wondered whether one could do this, he answered sharply, "The question is which is to be the master—that's all." Few of us have the power that Humpty Dumpty claimed, however, and words as we use them are often vague or ambiguous.

Vagueness exists when it is not clear what a word or sentence means, when the meaning is fuzzy or cannot be fully grasped. Ambiguity exists when a word or sentence has two or more clear meanings but we do not know which of them to assign to this particular use. Vagueness and ambiguity are thus different, and they are overcome differently. Vagueness is eliminated by settling on a single meaning. Ambiguity is eliminated by listing the various alternative meanings and choosing the one that fits the situation.

How can we settle on a single meaning for a vague term or sentence? It is not a matter of arbitrary decision, for knowledge of matters of fact is important. When we say, "It is hot," we are using a very vague word, "hot." If we want to speak more exactly, we will have to know what "it" refers to. If "it" is the weather, we may mean it is too warm for human comfort, although even that may be more than we are entitled to say. People born and raised in the tropics may be quite comfortable at the same temperature. So the sentence may mean only "It is too warm for my comfort." But if "it" is soup, "hot" may mean "temperature high enough to burn the tongue." Again, if "it" is iron and the sentence is spoken in a smithy, "hot" may mean "warm enough to be malleable." The word "hot" refers to a continuum of temperatures from the slightest warmth to the most intense heat, and eliminating vagueness in use of the word depends on knowledge of the context, and of the purposes to be served.[2]

[2] Insofar as we are concerned with the intent of the speaker and the context of his utterances, this is an instance of pragmatics.

Finding the meaning of a vague sentence may turn it into an ambiguous sentence, because its very vagueness allows it to have several meanings. A usually lucid and exact writer, Anatole France, said, "The unknowable envelops and throttles us." What does the sentence mean? It may mean nothing at all and be only an expression of some feeling of M. France in the face of the complexities of existence. If it does mean something, two meanings suit it equally well: much is beyond human reason, and the attempt to understand it thwarts and frustrates us; or, there is some agent in the universe that holds us back from ultimate knowledge, perhaps knowledge of itself. Converting this vague sentence into an ambiguous one allows us to see which of the two meanings best suits the context in which the sentence occurs. That is a step forward, but it confronts us with a difficulty that arises when we argue about ambiguous sentences: confusion between real and verbal issues.

Consider this story by William James.[3]

> Some years ago, being with a camping party in the mountains, I returned from a solitary ramble to find everyone engaged in a ferocious metaphysical dispute. The *corpus* of the dispute was a squirrel—a live squirrel supposed to be clinging to one side of a tree trunk; while over against the tree's opposite side a human being was imagined to stand. This human witness tries to get sight of the squirrel by moving rapidly round the tree, but no matter how fast he goes, the squirrel moves as fast in the opposite direction, and always keeps the tree between himself and the man, so that never a glimpse of him is caught. The resultant metaphysical problem now is this: *Does the man go round the squirrel or not?* He goes round the tree, sure enough, and the squirrel is on the tree; but does he go round the squirrel? In the unlimited leisure of the wilderness discussion had been worn threadbare. Everyone had taken sides and was obstinate; and the numbers on both sides were even. Each side, when I appeared, therefore appealed to me to make it a majority. Mindful of the scholastic adage that whenever you meet a contradiction you must make a distinction, I immediately thought and found one, as follows: "Which party is right," I said, "depends on what you *practically mean* by 'going round' the squirrel. If you mean passing from the north of him to the east, then to the south, then to the west, and then to the north of him again, obviously the man does go round him, for he occupies these successive positions. But if on the contrary you mean being first in front of him, then on the right of him, then behind him,

[3] *Selected Papers on Philosophy,* Everyman Ed., pp. 198 ff.

then on his left, and finally in front again, it is quite obvious that the man fails to go round him for by compensating movements the squirrel makes, he keeps his belly turned towards the man all the time, and his back turned away. Make the distinction, and there is no occasion for any further dispute. You are both right and both wrong, according as you conceive the verb 'to go round' in one practical fashion or the other."

Although one or two of the hotter disputants called my speech a shuffling evasion, saying they wanted no quibbling or scholastic hairsplitting, but meant just plain honest English "round," the majority seemed to think that the distinction had assuaged the dispute.

James understood that this was not a dispute about a real, or factual issue, but about a verbal one. The quarrel was about the meaning of the words "go round," which were used ambiguously. Perhaps one word in James's account should be changed, to make the matter clear. He wanted to follow the adage, he said, "that whenever you meet a contradiction you must make a distinction." But a real *contradiction* exists when one of two propositions is true and the other false, or when the same proposition is affirmed and denied. The word that should be changed is "contradiction," which should become "paradox." A paradox is a seeming, not a real, contradiction, and it may be resolved by showing that, due to an ambiguity, *different* propositions are being affirmed and denied, although they seem to be the same. They seem to be the same because the same words are used, but they are really different because at least *one* word is used to mean one thing in one statement and another thing in another statement. So the statements, although their words are the same, may really be two different and quite independent propositions. And, instead of contradicting each other, they may both be true or may both be false.

In the problem of the squirrel, the ambiguity is resolved by distinguishing two meanings of "go round." The first meaning of "go round" (the squirrel) is "passing from the north of him to the east, then to the south, then to the west, and then to the north of him again" and the second meaning is "being first in front of him, then on the right of him, then behind him, then on his left, and finally in front again." The question "Does the man go round the squirrel or not?" is now two questions: (1) Does the man pass from the north of him to the east, etc.? (2) Does the man move from being first in front of him, then on the right of him, etc.? The answer to question number one is yes, and the answer to question number two is no.

The "quarrel" was no quarrel at all; the two propositions are separate and distinct.

A verbal issue, like this one, often seems to be a real issue, as this did. One should perhaps speak of "factual" issues instead of "real" ones, for verbal issues are not in themselves spurious. What is spurious is passing off as a factual issue one that is verbal. The distinction is this: a factual, or "real," question is about the truth of a proposition, and a verbal question is about the meaning of a proposition. And, obviously, one should first know what a proposition means before he asks if it is true.

Sometimes, though, spuriousness is the other way round. A seemingly verbal debate may mask a factual or real issue. The question is not always, "In which sense do I mean 'going round'?" It may be, "Which sense of 'going round' is most relevant to my purposes?" or "Which sense suits the context, or allows my problem to be solved?" If a squirrel hunter kept moving around a tree and the squirrel kept the trunk between itself and the hunter, it would be relevant to tell the man that the squirrel was still on the tree but that it was moving so that the hunter was not "going round" it. But it would be irrelevant to tell the hunter that he was "going round" the squirrel in that he was moving successively north, east, south, and west of the squirrel. For the former statement might help get the animal in the sights of the rifle, and the latter has nothing to do with the case.

Language is usually metaphorical—as thought is—and without some knowledge of the rhetorical and poetic uses of words we remain simple-minded about meanings. A metaphor explains, or clarifies, the meaning of a term or a relation by comparing it with another term or relation.[4] The comparison is with something like, or identical to, the original term or relation in a particular respect. So if I say, "The meeting was a Tower of Babel," I may be trying to show how totally the people who were present failed to understand each other.

A *simile* may be defined as a metaphor containing "like" or "as" or an equivalent. ("The meeting was *like* a Tower of Babel.") But a metaphor may also be thought of as a compressed simile. In "The

[4] Some novices have a tendency to use metaphor as sheer ornament, probably because they believe that an extensive use of metaphor is necessary for good style. But metaphor is useless, sometimes even a hindrance, unless it clarifies what is being said or at least makes the meaning more vivid.

meeting *was* a Tower of Babel," the compression of the metaphor is very slight, being only one word less than the simile, but in other instances the compression is considerable. A trite phrase, "his steely eye," is a metaphor probably compressed from the simile "his eye is as hard, gray, and cold as steel." Even this simile is metaphoric in its use of the words "hard" and "cold," for the eye is literally neither.

It is not uncommon to have a metaphor within a simile, for metaphor is almost indispensable in writing or talking; it can be minimized only by the use of technical terms, and avoided totally only by the use of mathematics. It takes a trained eye to find the metaphors lurking in almost every paragraph we read, but they are there. One reason for the pervasiveness of metaphor is probably the inadequacy of language to express literal meanings; another, much more important, is the nature of thought itself. We use metaphor because we see relationships of similarity and identity in the universe; if we did not, it might be impossible to put discrete ideas together into a coherent whole. The assimilation of new ideas by means of metaphor is especially valuable because it is much easier to understand them if we can see that they are in some ways like more familiar ideas. Thomas Hobbes said that kings stand to each other as men do in the state of nature, thus explaining that the relation between sovereigns is the same as the relation between individual men who live outside political society. The simile clarifies the idea of sovereignty as an ultimate authority and implies that it is the absence of a larger power to which kings owe obligation that is the reason for the conflict among them.[5]

[5] Perhaps we can sharpen the meaning of metaphor by using a simile: A metaphor is like a mathematical ratio. If we say that 4 is to 6 as 6 is to 9, we thus identify the first relation as "two-thirds of": 4 is two-thirds of 6 and 6 is two-thirds of 9. However, this is not the only possible relation between 4 and 6; if we wanted to call attention to a different relation between those numbers we could construct another ratio: 4 is to 6 as 6 is to 8, and so identify the first relation as being "two less than." In Hobbes's simile, the relation between kings to which he wanted to call attention (there are, after all, a large number of possible relations between kings) is their independence of a larger political context; and he made his point without stating it explicitly. Even if he had stated it explicitly, but in an obscure or difficult way, a metaphor in addition to his exposition would have been helpful.

A metaphor, valuable as it often is, is not proof of any assertion. Political campaigns, for example, are replete with arguments like "Don't elect a new candidate in this time of crisis. You shouldn't change horses in midstream." Perhaps not, but a crisis isn't a stream and a candidate isn't, actually, a horse. *The basic error is to interpret a metaphor or any other trope literally.* If we think of "falling from grace" or "falling into error" literally, we will probably expect "sinking" sensations when we do either. Unfortunately, there is no such immediate warning. And much worse is the literal interpretation of metaphoric statement in law and religion. Cato the Censor, great Roman puritan reformer, interpreted the old Roman religious law that on festival days "the plow should not work" as meaning that slaves must not use the plow on those days *but must continue to work at other tasks.* The Inquisition strictly obeyed an injunction that the Church should not shed blood by using tortures like the rack and the wheel and by turning condemned heretics and witches over to the secular arm for burning.

The most common metaphors are the hardest to recognize because they are single words, usually adjectival. There are "threatening" clouds, "raging" storms, and "peaceful" skies, "dead" languages, "live" options, and "sick" nations. All these are metaphors of men or animals used to characterize inanimate things. They are fairly unequivocal metaphors because they use a particular quality which has one obvious meaning in the context—although "sick" nations is not so clear as the others. But the "steely eye" which we mentioned earlier is equivocal as compared to a simile because a simile might call attention to a similarity in hardness or color, but not both. When we talk of the "body politic" we are using a metaphor, although not a clear one, which compares society as a political order to an organic body. Metaphors which do not use an unequivocal quality, like "raging," but compare two complex things, do not show *in what respects a* is like *b*—in this case how the political order is like a body. Such metaphors mislead when it is assumed that because two things are alike in one respect they are alike in others. James I, in his elaborate defense of the divine right of kings to rule, used the metaphor of the body politic in a sophistic fashion. If the political order is like a body, he said, then the king is like the head; if you remove a body's head, the body dies; therefore, if you depose the king, the political order is destroyed. Of course, the respects in which the state is like a body and its ruler like a head do not include this particular relationship of life and death. One cannot decapitate a man and substitute another head without killing him, but rulers can succeed each other without necessarily injuring the state.

It is equally wrong to suppose that if two things are different in some respects they must be different in all. So an analogy is not destroyed by showing that it breaks down at some other point. An analogy compares things in some particular respect in which they are alike. Speaking strictly, it compares them in terms of their relations or attendant circumstances; a trance is like death, for example, in that overt bodily motion ceases. And every analogy compares things that are in many ways dissimilar; otherwise there would be no point in it. Thus we question an analogy only by asking whether it holds in the one respect in which the two things are compared.

The king's argument above is an instance of reasoning by analogy. Reasoning by analogy is a very elaborate form of comparison in which an argument is explained, clarified, or made more vivid by showing its similarity to, or identity with, another argument in respect to its logical form. Sometimes, no matter how precise the statement of an argument, it does not carry conviction until a similar argument (similar, that is, in logical form) is made about another and more familiar matter. Suppose a university administrator were to argue that he wanted to employ a Nazi or a Communist to teach a course in totalitarian government because otherwise it would not be explained fully and well. He might try to justify his decision by pointing out that since it is proper to employ chemists to teach chemistry and philosophers to teach philosophy, it should be proper to employ Communists to teach communism and Nazis to teach Nazism. The analogy, of course, is a false one. We could show this by pointing out that the forms of argument he used only seemed to be, but were not in fact, the same. That a chemist should teach chemistry and a philosopher philosophy is a belief based on the principle that only people who are competent in a subject should teach it; and the word "chemist" means a man who is professionally competent in the subject of chemistry. But the word "Communist" only means a man who believes in or practices the doctrines of communism, perhaps in the sense of obeying the orders of Communist officials; there is no implication that he understands, can analyze, or explain those doctrines. Secondly, apprenticeship is involved in one but not the other. Some students will learn to practice chemistry by studying under a chemist, but we do not want students to practice communism vocationally. The analogies of the university administrator are, therefore, misleading.

Having explained why this is a false analogy, we might feel that we still had not persuaded our opponent. So we might resort to an analogy of our own by saying that to insist on employing a totalitarian to teach totalitarianism is like insisting on employing a criminal to teach criminology or a drug addict to teach about drug addiction. The absurdity of the original argument is now clear because a really analogous argument (one based on the same principle) is obviously absurd.[6]

There is an argument by analogy made by St. John Chrysostom: [7]

> I hear many cry when deplorable excesses happen, "Would there were no wine!" Oh, folly! Oh, madness! It is the wine that causes this abuse? No. . . . If you say "Would there were no wine" because of the drunkards, then you must say, going on by degrees, "Would there were no steel," because of the murderers, "Would there were no night," because of the thieves, "Would there were no light," because of the informers, and "Would there were no women," because of adultery.

[6] On the question of communists as teachers, see Sidney Hook, *Heresy Yes, Conspiracy No,* John Day, 1953, and Ernest van den Haag, "Academic Freedom and Its Defense," in *Strengthening Education at All Levels,* American Council on Education, 1953.

[7] Quoted in Max Black's *Critical Thinking,* Prentice-Hall, 1946, p. 40.

# 3

# Logical Analysis

We can predicate truth of a complete statement only, not of words alone. But if the thought contains no words with actual referents, yet has meaning, we do not talk of truth, but of validity.

The most systematic example of symbols that have no referent at all in sense experience is mathematics. Bertrand Russell once defined mathematics as a subject in which no one knows what he is talking about, or whether anything he says is true. The humor of this definition lies in the confusion of colloquial phrases with exact statements which mean something very different; one does not know what he is talking *about* in pure mathematics because the symbols in which he expresses himself do not stand for any thing or process. And what the mathematician says is, strictly speaking, valid or invalid—consistent with his premises, or inconsistent—not true or false.

The number 2 does not stand for pairs of actual objects,[1] although it is most readily illustrated by such reference. Like all numbers, 2 is a concept which may be defined entirely within the number system or defined so as to apply to things. But its application is different from what it is in itself. We *add* numbers, but we *combine* objects. In arithmetic, $2 + 2$ always equals 4, and we can apply this simple equation to things and get such correct statements as: two apples and two apples equal four apples. But if we combine two gallons of water with two gallons of alcohol we do not get four gallons of liquid because of the way in which water and alcohol mix. Arithmetic is still applicable. It

[1] Whether man invented numbers as a result of particular observations we do not know. But such speculation is irrelevant here, for we are concerned with what numbers signify, not how they originated.

tells us that whenever we combine two gallons of water with two gallons of alcohol we *have combined* four gallons of liquid. But the quantitative *result* of the combination cannot be determined by arithmetic alone, for it depends on the chemical properties of the two liquids and their reactions when mixed with each other. This is a factual, not a mathematical matter, and our knowledge about it never possesses certainty, as mathematics does. Although all alcohol is alike in some respects, we cannot be *absolutely certain* that every time we encounter alcohol it will be *exactly* as it was in the past—as so many drinkers have learned to their sorrow. But the number 2 is always the same; it has no physical, and so no changeable, existence. Two and 2 and the sound we make in saying the word are only symbols for a concept. And 2 in any of its uses is identical with 2 in all other uses. Therefore, whatever is true of 2 is true whenever we use the number; $2 + 2$ (as distinct from 2 of these objects and 2 of those) is always 4.

We say that $2 + 2 = 4$ is "necessarily true," using that phrase to stand for any statement the denial of which leads to self-contradiction. Since 2 is $1 + 1$, $2 + 2$ is $(1 + 1) + (1 + 1)$; and 4 is $1 + 1 + 1 + 1$. Remove the parentheses and it is clear that $2 + 2 = 4$ is the same as $1 + 1 + 1 + 1 = 1 + 1 + 1 + 1$. There are identical quantities on both sides of the equtaion, or, put differently, there is the same *number* on both sides of the equation. Therefore, to deny that $2 + 2 = 4$ is to deny that a number is the same as itself—a flat contradiction.

To summarize our point: Mathematics is a system whose symbols are manipulated in accordance with specified rules. All its valid conclusions are called "necessary," or "certain," because they cannot be other than they are without contradiction. Pure mathematics, as distinct from applied, or empirical, mathematics, contains no information about anything in the world of sense experience and is only the result of applying rules of procedure to definitions. When mathematics is applied to existent things and events, the results are no longer certain. For then the statements purport to convey information about things that can be experienced, and they are to be judged from the standpoint of truth, not merely that of validity. And empirical truths are not necessary, for they can be denied without self-contradiction. To deny that grass is green may be false, but it is not self-contradictory.

Mathematics in some ways is more like art than like science. Its qualities of imagination, precise organization, and relation of part to whole are essentially aesthetic. But its results, unlike art, are valid or invalid in a strict sense, and in application it is not only indispensable to practical pursuits like accounting and engineering—it is perhaps the greatest of scientific tools. Science can deal with exact quantities only by applying mathematics; and exactness is an all-important scientific virtue. It is an ideal toward which science moves, and if there were a final body of scientific knowledge the statements composing it would be exact. But exactness cannot be expected at every stage of scientific inquiry, and although we may demand as much precision as the state of knowledge and the nature of the subject matter allow, we must be careful that we do not sacrifice imprecise but rich insights, pregnant with scientific promise, for an arid precision that measures to no real purpose and stops more important investigation. This is a caution especially important for social science, which is sometimes pushed too fast to the model of the more honorific natural sciences.

Even in natural science, the example of two gallons of alcohol and two gallons of water should warn us that we cannot combine objects without observing the way they act in relation to each other, nor can we combine gases or liquids without consideration of the effects of temperature and pressure on their volume. But with simple precaution in its use, mathematics can be applied with the utmost fruitfulness.

Pure logic has the same general characteristics as mathematics.[2] Like mathematics, logic is a set of rules for attaining correct conclusions from premises; its terms are abstract and its conclusions are valid, not true. So it is not the content, if any, but the *form* of logic (and mathematics) which determines its validity. This is what is meant when we call logic and mathematics "*formal* sciences."

When particular words which are interpreted as having referents in sense experience are substituted for logical symbols, conclusions which convey information are attained, and they can be true or false. If all *A*'s are included in *C*, and all *B*'s are included in *A*, than all *B*'s are included in *C*. This is a correct form of reasoning, and valid, but *A*, *B*, and *C* are symbols of great generality. When we apply this reasoning to anything empirical, we substitute words for abstract symbols and we get such a statement as the classic syllogism: "If all men are mortal, and Socrates is a man, then Socrates is mortal." This is

[2] Logic is sometimes treated as the basis of mathematics, and sometimes as a branch of mathematics which employs a different set of symbols. A discussion of applied logic may be found in the section "Analytic propositions and empirical propositions."

still only a valid, not a true argument, an instance of the fully generalized syllogism, because of the word "if." No assertion was made that anything was so. The syllogism means that *if* what is said in the premises is true, *then* what is said in the conclusion is true. When we drop the "if," we are asserting that the premises *are* true and that the conclusion, therefore, is true. Although the conclusion still follows *validly* from the premises, it may not be *true* because the premises may not be true; and logic alone cannot determine whether or not *in fact* the premises are true. That is the task of empirical science.

In a purely logical argument we use symbols of the most general kind (the *A*'s, *B*'s and *C*'s of the last paragraph) in order that the form may be applied as widely as possible. These symbols are called "variables," and the statements in which they appear are called "propositional functions." When we substitute a particular word for a variable, that word has, presumably, one and only one meaning as we employ it, so we call it a "constant." When all the variables in a propositional function have constants substituted for them, the propositional function has become a proposition, an actual assertion or denial, capable of being either true or false.[3]

Nothing but propositions can be true or false. *Things* may exist or not exist, be here or there, but they are not true or false. *Concepts* can be clear or fuzzy, fruitful or sterile, but they cannot be true or false, for they neither assert nor deny. If a statement is ambiguous it contains two or more propositions; if it is vague it may contain none, although perhaps suggesting many. Propositions, therefore, cannot always be recognized by the form in which they usually appear, that of sentences, but are identified when it is clear that something is being asserted or denied. "Pyrrhus the Romans will defeat," said the ancient oracle, playing safe by stating two contradictory propositions in one sentence, and so not clearly asserting or denying either one.

A *proposition* is the logical aspect of a statement, its meaning. A

[3] As the word "variable" is employed in science, it often stands for any general term that is relevant to a problem and has particular applications. If we say, "A man's social status is determined by such matters as income, occupation, and education," the words "income," "occupation," and "education" are variables, for we do not know what class status to assign to Mr. Jones until we replace the general terms with particular ones, like $8,500 per annum, barber, and high school graduate.

*sentence* is the grammatical aspect of a statement, an ordering of words which in English has a subject and a predicate. A *judgment* is the unexpressed idea in our minds which perhaps precedes a statement. So we make judgments (psychological) and utter sentences (linguistic and grammatical) which convey propositions (logical).[4] Different names are used for types of propositions and sentences; thus a proposition that makes an unqualified assertion or denial is a *categorical* proposition, but the sentence that states it is a *declarative* sentence. The meaning of a *conditional* sentence is a *conditional,* or *hypothetical,* proposition, and is usually stated, for purposes of logic, as "If . . . , then. . . ." All scientific laws are hypothetical.

Consider the following: it is a horse; that object is a horse; *c'est un cheval; ese es un caballo; das ist ein Pferd.* From the standpoint of grammar, these are five sentences in four languages; from the standpoint of logic, these are a single proposition because their meaning is the same. Grammatically, these are simple declarative sentences containing subject, verb, and predicate. Logically, there is one categorical proposition containing two terms and a relation.

Logic normally distinguishes propositions into three kinds in terms of what is called "quantity," that is, how many members of a class of objects are referred to by the subject. The three kinds of quantity are: (1) general, or universal; (2) particular; (3) singular, or individual. General propositions have as their subjects all members of a class (*all men, men, man* are general subjects); particular propositions have as their subjects an unspecified number of members of a class (*some men, many men, most men* are particular subjects, as are any other terms that mean *at least one*); singular propositions have as their subjects *only one* member of a class (*John, this man, that man* are singular subjects). It follows that "all men are mortal" is general, or universal; "some men are mortal" is particular; and "Socrates is mortal" is singular, or individual.

Scientific laws are general, or universal propositions. Yet at the same time they are, as we said, hypothetical. This is so because we

[4] A distinction like that between proposition and judgment exists between implication and inference. Implication is a logical relation between propositions such that if one of them is true, then the other is true. Inference is the temporal process whereby *we* reason from the truth of one proposition to the truth of another. We infer correctly that proposition $q$ follows from proposition $p$ when $p$ logically implies $q$.

interpret a general categorical proposition as hypothetical. Thus "all men are mortal" becomes "if there is any $x$ such that $x$ is a man, then $x$ is mortal." General propositions do not imply the existence of their subject; they assert that *if* instances of the subject do exist, then what is predicated of the subject is predicated of those instances. Particular propositions, however, are interpreted as implying existence. Thus "some men are mortal" becomes "there is at least one $x$ such that $x$ is a man and $x$ is mortal." [5] These logical distinctions are necessary for an understanding of science.

Traditionally, the logic of discovery, or the procedure of science, has been called "induction" and the procedure of logic and mathematics has been called "deduction." However, the old distinction between these two is faulty. Deduction, it was thought, was inference from the general to the particular, and induction was inference from the particular to the general. But the reasoning involved in induction is not so distinct from deduction as this makes it seem. One syllogistic rule is that there must be at least one—but not necessarily more—general, or universal, premise. For particular premises alone do not yield a conclusion. From the premises "Some apples are red" and "Some of these objects before me are apples," for example, there is no conclusion. "Some of these objects are red" does not necessarily follow from the premises; those apples which are red may be other apples than the ones I have before me.

Induction, according to the old definition, worked like this: If I were to look at a crow and see that it was black, then to look at a second crow and see that it too was black, then a third crow, a fourth, and so on, until I had examined a large number of crows, I would conclude

[5] There is a rule of the syllogism, the classical model of deduction, that in order for a conclusion to be particular, *one* of the premises must be particular. So it is valid to argue:

| | |
|---|---|
| MAJOR PREMISE: | All men are mortal. |
| MINOR PREMISE: | Some Americans are men. |
| CONCLUSION: | Therefore, some Americans are mortal. |

The same conclusion which implicitly asserts that at least one American exists would not follow validly if the minor premise said "all" instead of "some." For then neither of the premises would assert the existence of any Americans.

that all crows are black. Of course, there is no assurance that the very next crow to be observed would not be of a different color. I may be examining the crows of a single country or continent and these may be black, whereas crows elsewhere are multicolored. But this, it was pointed out, was the necessary risk of science.

Actually, the risk is much greater than science need take because something has been left out. We do not merely assume, for no reason, that the crows we have examined are representative of all crows with respect to color. Rather we follow rules for the selection of samples which give a high probability that they are fairly representative of all members of their class, in this case of all crows. Now, this belief that the particular crows we have observed are representative is a general proposition: "All crows have the color of these crows." Our induction, therefore, makes logical sense only by including a general proposition implicitly. So science should not be thought of as purely inductive, in the old sense. Its method contains deduction and may be called, as a whole, "hypothetico-deductive."

The principles of logic are basic to science, as they are to all thought. And basic to logic are the so-called "laws of thought," which are necessary conditions for valid thinking. These laws define logical consistency.

THE LAW OF IDENTITY: If anything is *A*, it is *A*.

THE LAW OF CONTRADICTION: Nothing can be both *A* and not *A*.

THE LAW OF EXCLUDED MIDDLE: Anything must be either *A* or not *A*.

*A* can be anything at all. If *A* is a pig, then identity asserts that anything that is a pig, is a pig (with which few would want to quarrel); contradiction asserts that nothing that is a pig is at the same time not a pig; and excluded middle asserts that everything you can think of either is a pig or is not a pig. The point is that we divide the entire universe into two mutually exclusive parts whenever we think about contradictories.

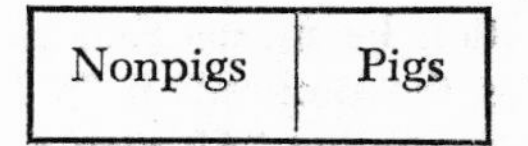


The idea of nonpigs, all things that are not pigs, includes everything conceivable except pigs. So, once we have divided the universe

in this way, there is nothing in it beyond the two parts. And that is what excluded middle states, for there is no middle ground or third part in which things may exist. Identity states that a thing is in whichever of the two parts it is in, and contradiction states that, since the parts are mutually exclusive, a thing cannot be in both parts at once. Put another way, contradiction states that anything that is a pig is not a nonpig, and anything that is a nonpig is not a pig.

The three laws may be applied to the truth-value of propositions if the word "false" means "not true" and the word "true" means "not false." There are, then, no other possibilities but these: (1) a proposition is either true or false (excluded middle), (2) a proposition which is true is true and a proposition which is false is false (identity), and (3) no proposition can be both true and false (contradiction).

One objection to these laws comes to mind at once. What about change? Of course, at any one time things are what they are and not something else, but over a period of time they change and do become something else; living men become dead ones, water becomes vapor or ice, and so on. The answer is that the laws of thought do not deny change, nor even mention it. They are principles of intelligible and correct thinking, and they always work when two things are assumed: *at the same time*, and *in the same sense*. These two assumptions may be added to each of the laws when it is being stated.

Literally, *A* and non-*A*, pigs and nonpigs, are not contradictories; they are opposites. Contradictories are the propositions which assert or deny that something is *A*, or something is a pig. And the laws hold of these propositions. One of the foundations of logical thought is the meaning of contradictory propositions. We will state this meaning in terms of the mutual exclusion of the words "true" and "false."

Two propositions are called contradictories when both cannot be true and yet both cannot be false; if one is true the other is false, and if one is false the other is true. Perhaps the most obvious instance of contradictories occurs when a proposition is asserted and then the same proposition is denied. If I state a proposition, *p* (we assume that when anyone states a proposition he means that that proposition is true), and then someone states the proposition "*p* is false," he has contradicted me. The way to contradict "All men are selfish" is not by "No men are selfish," for both of those propositions may be false (if, for example, the truth is that some men are selfish and some are not). The contradictory of "All men are selfish" is "Some (i.e., one or more) men are not selfish." If either of these propositions is true, the other is false, and if either is false, the other is true. This is a matter of logical analysis. But which proposition is true and which is false is not a matter of logic but of fact.

The scholarly habit of too great caution, of refusing to commit oneself, saying, "The truth is somewhere in the middle," or "I cannot fully accept either alternative," becomes preposterous when it is used to apply to contradictories. As the law of excluded middle shows, there is no middle. One or the other proposition must be true.

If we cannot find a meaning for the propositions that we hear, or if we deny that the meaning is clear, then we are really denying that they are propositions at all, and, of course, thus denying that they are contradictories. We may also refuse to accept the problem that the speakers think they are answering. But if we accept the problem and know what the answers mean, then there is no alternative proposition that can be true. If one man asserts, "There is a destiny that shapes our ends," and another says, "It is false that there is a destiny that shapes our ends," then, if we think we know what they mean and do not wish to suspend judgment, we must agree with one and we cannot agree with both.

It is fatal to thought to confuse contradictories with contraries. Two propositions are called contraries when both cannot be true but both can be false; i.e., if one is true then the other is false, but if one is false then the truth value of the other is logically undetermined. We have said that "No men are selfish" is not the contradictory of "All men are selfish." It is the contrary. Both propositions cannot be true but it is possible that both are false (as in our previous example, if some men are selfish and some are not).

The chief error in dealing with contraries is to treat them as contradictories. The history of thought is full of instances in which people have believed that they must choose one of two beliefs when, actually, those beliefs have been contraries and other alternatives have been possible. Thus, the need felt by many thinkers to decide whether the world was *essentially* matter or spirit neglected the possibility that it was neither. One way of proving the truth of a proposition is to prove the falsehood of its contradictory. This does not work with contraries since some other, perhaps as yet unformulated, proposition may be the true one. Thus, to give up the belief that human action is governed by calculated self-interest does not warrant accepting the belief that man's behavior is always unreflective. In recent years we have witnessed a growing disbelief in the notion that all man's problems can be solved rationally. Unfortunately, many people conclude from this that we should give up the attempt to solve problems rationally. What they have done, of course, is to assume incorrectly that "Reason can solve all problems" and "Reason can solve no problems" are contradictories. Actually, they are contraries. The truth may be that reason can solve some problems and not others. It may even be correct to say that although reason cannot solve all problems, nothing but reason can solve any problems.

In a dispute between a monotheist and a polytheist, a successful attack on one position does not necessarily prove the other. They are contraries, not contradictories. A third position is possible—that of the atheist. Equally, to say "Man's life is entirely a comedy" does not contradict the assertion "Man's life is entirely a tragedy," although it denies its truth. Both may be false and other alternatives may be possible. But when a man says, "If wages rise, then within six months prices will rise," and a second man says, "Wages did rise over six months ago and prices did not rise," then the second man has contradicted the first; one is right and the other is wrong.

The errors arising from violation of these laws are only a few of the many fallacies which bedevil thought. Let us now consider a few others, major ones. The names of the fallacies and the Latin tags associated with some of them may seem like mere pedantry, but it is valuable to know them because they are used so often in books of the most varied kinds. The examples are oversimple because they have been selected for pedagogical purposes, but they are not trivial, for they do carry the weight of the general principles involved. And they are short enough to be remembered, so they can be used as touchstones to reveal fallacies in more complex cases.

Fallacies are usually divided into three types: *formal, verbal,* and *material.* Formal fallacies are violations of the formal rules of logic, such as the three laws of thought and the rules of the syllogism. Verbal fallacies arise from confusion of words and grammar. Material fallacies are fallacies of argument that are not actual violations of logic.

We will not discuss formal fallacies because we have not covered logical rules, and what constitutes violation of the laws of thought is obvious from what has already been said. Some verbal fallacies were discussed earlier under the headings of *ambiguity* and *vagueness*. Technically, ambiguity of individual words is called *equivocation*, and ambiguity due to sentence structure is called *amphibology*. The prophecy in Shakespeare's *Henry VI* is an instance of amphibology: "The Duke yet lives that Henry shall depose"; so are the imaginary advertisements dear to the hearts of schoolboys: "Wanted: a piano by a lady with carved legs," and "a watch by a man with an open face."

Three more verbal fallacies should be mentioned, the first two of which are special cases of equivocation. The *fallacy of composition* arises from confusion of a universal and a collective term. A universal term stands for all its instances separately. When we say, "All objects in our atmosphere tend to fall to earth," we are asserting something of each and every object taken separately. But when we say, "Crowds tend to panic in an emergency," we do not mean that each man in the crowd taken separately tends to panic in an emergency. The fallacy of composition consists in concluding something about a universal term as if it were collective. So it is composition to argue: "Every man on this ball team is under thirty-five years of age; therefore this ball team is under thirty-five years of age."

The *fallacy of division* is the converse of composition. It infers something universal—hence individual—about a collective term. So it is division to argue: "The United Nations has no great military force; the United States is one of the United Nations; therefore the United States has no great military force."

The *fallacy of accent* consists in changing the meaning of a sentence by emphasizing or accenting the wrong words. The logician and economist W. Stanley Jevons gives an amusing instance of accent which may occur in reading Chapter 13, verse 27, of the First Book of Kings. The quotation is "And he spake to his sons, saying, Saddle me the ass. And they saddled *him*." The word is italicized, which means that it was supplied by the translators. But an unwary reader might think that the emphasis was supposed to fall on the word in italics. Quite different meanings result from emphasizing in turn each word in even the simplest sentence. Thus "Give me the book" has four separate meanings as each of its words is accented, and none of them, probably, is the meaning of the sentence when no particular word is emphasized.

There are five very common material fallacies. The *argumentum ad hominem* denies the logic or truth of an argument by abusing the man who advances it. What a scoundrel says may be true or logically valid, and it is no refutation of his statement to prove he is a scoundrel. This fallacy is employed frequently, though seldom deliberately, by people who have contracted the modern disease of psychologizing at all times, whatever the issue. So it is an *argumentum ad hominem* to answer a question about the logic, truth, or relevance of a man's assertion by attributing unpleasant motives to him.

*Begging the question,* or *arguing in a circle,*[6] consists in using as part of the evidence for itself the conclusion that is to be established. It is like being asked for a reference and naming a Mr. Smith; then, since Mr. Smith is quite unknown, saying, "Oh, but *I* vouch for him." This may seem absurd, but we have often heard arguments like this: The first five books of the Bible were written by Moses and must be true because Moses was inspired by the Lord; and the fact of Moses' divine inspiration is evident because the Bible tells us so. There are also many pseudo-scientific arguments that are circular. When we are told that in the struggles of society the fittest survive, and we ask, "Fittest for what?" we may be answered, "For survival, of course." Jeremy Bentham spoke of "Question-begging Epithets," which are single words that can beg a question. The word "enough" is often used in this way. When we are told that a man who has enough talent will rise to the top in any society and we ask how much talent is "enough talent," it may turn out that a man is only said to have enough talent when he does in fact rise to the top.

We need only mention *non sequitur* in which, no matter how showy or elaborate the argument, the conclusion simply does not follow from it. But more deceptive is the fallacy known as *post hoc ergo propter hoc* (after this, therefore because of this) in which it is assumed that because *A* precedes *B, A* is the cause of *B.* Countless superstitions are instances of this fallacy. If there are shooting stars the night before a great man's death, there will be people to argue that the heavens showed omens, or that man's life is connected with the stars and when one's particular star falls, he dies.

Paracelsus, physician-magician of the Renaissance, had a scientific triumph with a use of magic that was an instance of *post hoc ergo propter hoc.* He revived a medieval superstition called "weapon oint-

[6] Known technically as *petitio principii.*

ment," the belief that instead of applying salves to wounds they should be applied to the weapons responsible. "Weapon ointment" promptly saved many lives on the battlefield, and the new cure was used by other physicians. What in fact happened was that many wounds healed naturally, whereas had the salve been applied they probably would have become gangrenous—for the salve contained substances as little antiseptic as cow's dung and viper fat.

Finally, there is the *fallacy of false question* (or of *many questions*), in which several questions are asked as though they were one. Such a question cannot be answered simply without assuming the answers to questions that have not even been considered. The old joke "Have you stopped beating your wife?" is a false question. Whether you answer yes or no, the answer assumes that you have previously been asked, "Do you beat your wife?" and have answered, "Yes." When the billboard in front of a church announces that the Reverend Mr. Jones will preach a sermon on the question "What Is the Purpose of Human Life?" it is assumed that a prior question, "Does Human Life Have a Purpose?" has already been considered and answered affirmatively.

The *genetic fallacy* does not fit any of our rubrics exactly. But we will describe it here, although it is sometimes listed as an error in scientific method, not logic. Essentially, it is the confusion of logic or science or values with history or genesis. The genetic fallacy is committed when we want to know if a proposition is true and are told how it came to be believed or uttered. And it is also committed when we ask about the origin or development of something and are told about its structure, its value, or its truth. An example of the former occurs when a judge condemns a criminal and someone asks whether the verdict is just, only to be told that the judge was in a bad temper that day because he quarreled with his wife at breakfast. Indeed, that may be *why* he decided as he did, but the issue of justice is one only of law and the merits of the case. Good temper or bad temper may be adduced as answers to the question why the judge said what he did, but not as answers to the question whether *what* he said was right.

Marxists often disparage the conclusions of social science by stating the conditions under which they came about. But everything comes about under some conditions, and to deprecate a belief because, say, economic conditions fostered it, is as foolish as to denounce the Ten Commandments because they were received at the top of a hill.

Equally, if we get a large check as a gift from a friend and wonder where he got the money—a historical question—it is no answer to be told the check is good.

The kind of reductionism we have decried also has an element of the genetic fallacy. When a man talks of the value of love, he may be told that love is only a biochemical state. But love may *appear* under biochemical conditions without being reducible to them—as science is not reducible to economic conditions—and no statement of conditions whatsoever is directly relevant to the *value* of the thing produced.

Fallacies are composed of meaningful statements, and we can eliminate most fallacies by altering an argument. But even that can not help when our statements are meaningless. The form of a sentence is usually the same whether or not it has any meaning, that is, whether or not it conveys an empirical proposition. If a sentence is obviously meaningless we do not know how to respond to it. "Fish are intended optimalities"—which has the same form as a meaningful sentence—would leave us confused, and we would respond by asking, "What do you mean?" But there are sentences that seem to have meaning—and to which we respond in a predictable manner—which, therefore, doubtless mean *something,* but do not yield that meaning to logical or semantic analyses of the sentences themselves. Thus if a political speech includes the statement "From the rockbound coast of Maine to the sunny shores of California, our party girds its loins, takes the bull by the horns, and prepares to run for a touchdown," the confusion of metaphors makes it impossible to know by analysis of the words, referents, and form of the sentence just what the sentence means. But we can be confident because of past experience that the speaker is urging: "Work for our party," and that the words he uses add only "Hurrah!" Even if he had stated this meaning explicitly, it would not be a proposition; exhortation enjoins action but does not assert or deny anything. We must distinguish, then, between all seemingly factual statements which turn out, on analysis, to be commands, suggestions, ejaculations, and the like, and those which are genuine propositions. The former we call "emotive" statements and the latter "informative" or "cognitive" statements.

To say that a proposition is false is not, then, the worst thing one can say about it. It is an intellectual accomplishment even to be wrong, because only meaningful statements can be false (or true). Meaningless statements cannot be either because there is nothing that they assert or deny. It is difficult to find an unequivocal criterion of factual meaning, a way of distinguishing emotive statements from informative ones, but there is one criterion in current use which is at least helpful.

According to the *verifiability criterion,* a statement is meaningful only if we can tell how it could, in principle, be verified; that is, if we can state under what conceivable conditions the statement's truth value could be tested. So the statement "That shaft is wider at the bottom than at the top" is meaningful because we know that the simple application of a tape rule would be sufficient to prove the assertion true or false. But the statement "The man who built that shaft was inspired by the spirits of his ancestors" is not meaningful in the same way. It may be a metaphoric statement, it may arouse respect or piety, but taken literally we do not know what human behavior could test it or what in sense experience is referred to by "the spirits of his ancestors," a phrase which *seems* to, but does not, stand for identifiable things.

The reason for the phrase "in principle" and the adjective "conceivable" in our formulation of the verifiability criterion of meaning is that there are many statements that cannot in fact be tested at the moment, or perhaps ever, which *could* nonetheless be tested *if* a specifiable set of conditions existed. The need for this qualification can be seen in an example offered commonly only a few years ago: "There are mountains on the other side of the moon." It was argued, by those who objected to words like "in principle" and "conceivable" in the verifiability criterion, that we see only one side of the moon from the surface of the earth and there is no way of observing the other side. The defenders of the qualification pointed out that, even if men could not in fact now observe the other side of the moon, there was nothing contradictory about the idea that they might one day be able to do so. Both sides seemed to regard it as dubious that man would ever see the other side of the moon. But even had man remained earthbound over the years, the point would have been the same. The meaning of the sentence, in terms of the ways of testing it, is perfectly clear. If an observer could be placed so that the other side of the moon were before him, then his observation could test the assertion that there are mountains.

A statement is testable in principle when the test itself is a conceivable one, even though it is not then physically possible. The "inconceivable" is not just a limitation in imagination, for some men can imagine things that other men cannot; "inconceivable" is used here to mean logical incompatibility or contradiction. Thus a test which would require observation of a circular square is inconceivable because "a circular square" is a phrase containing two incompatible ideas, a circle and a square. If the hypothesis about the other side of the moon were that there are mountains on it that are at once more and less than one hundred feet in height, it would not be testable in principle. We cannot conceive anything that is both more and less than one hundred feet high because it is a contradiction, so there would be nothing that could possibly be observed that would bear on the truth or falsity of the statement.

A statement is testable only if we can decide what possible observation would prove it true *and* what possible observation would prove it false. It is not enough to say, "If we can observe *A*, *B*, and *C* under conditions *y*, then a given statement is true." We must also say, "If we cannot observe *A*, *B*, and *C*, or if we observe non-*A*, non-*B*, and non-*C* under conditions *y*, then the statement is false." If, for example, I am indoors and I assert that it is raining outside at this moment, my statement is tested by looking out the window or going out the door. If I see or feel rain, and no one is pouring water from the roof, or some such thing, then I shall be justified in feeling that my assertion was correct. But if I see and feel no rain, then I must be willing to concede that my assertion was false.

It follows that a statement that seems true under *all possible* conditions is not empirically meaningful. If I say, "There is weather outside," my statement could not possibly be proved false, for whether it is raining or not raining, or whatever the day is like, there is still some kind of weather. More serious instances of statements which can under no *conceivable* conditions be shown to be false are predictions without reference to time. If I say, "There will be a violent revolution in the United States which will end in the establishment of socialism," it is easy to know what observations would show the statement to be true. But no conceivable happening could ever show it to be false because I can always say, "This situation is only temporary; wait, and you or your descendants will see the socialist revolution." People who make such arguments have the consolation that no one can ever show them to be wrong. But that is their weakness, not their strength.

Yet there is a sense in which we all know the meaning of "There will be a socialist revolution," and we know what it means to say, "I'm not sure when it will happen, but I am sure that it will." Who will deny that all men are mortal, though there are no conceivable conditions under which it could be refuted. If a man lived to be ninety thousand years old one could always say, "His age is no proof that he is not mortal; he still may die." Yet surely the statement that all men are mortal means something to us, something of the greatest importance, and the fact that it cannot in principle be tested does not rob it of significance; we might all behave differently if we did not know it to be true.

So there are other uses of the word "meaning," quite unlike those we have considered. "Does life have a meaning?" is sometimes regarded as asking whether life has purpose or value, not, literally, "meaning." But "meaning" can be construed quite literally in this question, for purpose or intent gives meaning. Words are not the only signs; artifacts—like flags and crosses—and actions—like rituals and dances—have meaning, and so does deliberate behavior. If I spend today doing things which will bring about an intended effect tomorrow, then the meaning of today's activities is tomorrow's occurrence. In general, the meaning of behavior is the intention that guides it, which is also the end it is calculated to produce. When someone acts in an unexpected way we ask, "What is the meaning of this?" An irate father who shouts this question at his daughter when he surprises her kissing a young man outside the door, may be answered, "We love each other," another way, perhaps, of saying, "We intend to be married."

Analogically, a man may say something cryptic or vague. When we ask what he means we are not asking what the statement means but what meaning he intends. We are answered satisfactorily by a clear statement which communicates what he intended to assert. When we follow a statement by "That is not quite what I meant," we are saying that the statement does not convey our intention.

The verifiability criterion of meaning, then, can be of immense aid to science. It forces us to render our meanings exact, and it connects the conceptual parts of science with the empirical parts in a single fabric. But it should not be used to hinder fruitful speculation even in science, and it is obviously too limited to encompass all of human thought. The attempt to apply it rigidly outside the sphere of science truncates imagination and narrows communication.

Some scientific philosophers in our century have tried to apply the verifiability criterion to every utterance and have concluded that whole areas of human thought and creativity, much of religion and art, for example, are meaningless. Thus they empty human life of significance. Perhaps it is best to think of the criterion either as a definition of *scientific* meaning, or simply as heuristic for the conduct of science. Then, if some theological or poetic statement does not satisfy the criterion, it can be said that the statement is not scientifically meaningful or that it is unacceptable in the body of science, but it may equally be said of a scientific statement conforming to the criterion that it is theologically or poetically meaningless, or outside the body of theology or poetry.

There is one type of proposition that is both meaningful and scientific yet nonetheless has no *empirical* meaning. To attempt to test it *empirically* is useless, yet it can be tested and is either true or false. It is utterly indispensable to science and to all thought. But to confuse it with empirical propositions—an easy mistake—is to lose all clarity. This type of proposition is called *analytic*.

All propositions may be classified either as analytic or empirical.[7] These are distinguished from each other by the *way* in which we can discover whether their propositions are true or false. Analytic propositions are those whose truth value can be tested by logical analysis alone. Empirical propositions are those whose truth value can be tested only by adding observation. Thus there are four kinds of propositions included under these headings: logically (or analytically) true, logically false, factually (or empirically) true, and factually false. Analytic propositions are logically—and so necessarily—true or false, but they contain no knowledge that is not implicit in the meanings of their terms.

These are instances of true analytic propositions: two plus two equals four; a normal horse has four legs; there are thirty-six inches in three feet; totally bald men have completely hairless scalps. To take

[7] We are omitting for the time statements that express value, like "*A* is more beautiful than *B*," or "*A* is morally better than *B*." These are called normative statements and there is controversy about whether they are reducible to analytic and empirical statements or are a separate class.

the last instance first: suppose you were asked to find out whether it is true or false. What could you observe that would be in any way helpful? If you collected a fair sample of totally bald men and examined them one at a time, what would you say if you discovered a few with hair? Obviously, you would not conclude that some totally bald men do indeed have hairy scalps. You would decide instead that they had got into the sample by mistake, because they were not totally bald. No conceivable observation, then, could aid you with your problem, for the meaning of "totally bald men" is identical with the meaning of "men with completely hairless scalps." So there was nothing to look for. You had only to analyze the meanings of the words in the sentence to discover that it was necessarily true. Denial of the sentence would imply a contradiction: that something is not the same as itself.

Equally, the meanings of "two," "plus," "equals," and "four" are, as we have seen, sufficient evidence that two plus two equals four. What is meant by "horse"—that it is a quadruped—and how we define "feet" and "inches" will be enough to enable us to know that the other sentences are true. True analytic statements, then, are tautologies; the meaning of the predicate is contained in the meaning of the subject. False analytic statements are contradictions: the meaning of the predicate contradicts the meaning of the subject. All the examples of analytic statements which we have given are tautologies. An example of a contradiction is: totally bald men have hairy scalps. Here the meaning of the predicate contradicts the meaning of the subject; by definition no man can be completely bald and have a hairy scalp.

Now for some instances of empirical statements: That tie is red. Mr. Smith weighs one hundred and eighty pounds. The Empire State Building is higher than the Chrysler Building. To test the truth value of these sentences, certain physical operations must be performed and observations made. To know how much Mr. Smith weighs, one must place him on a scale and look at the number to which the marker points. To test the other two statements, one must observe the color of the tie and the results of measurements of height.

The basic difference between the analytic and the empirical may be illustrated by the difference between a statement that Mr. Smith weighs one hundred and eighty pounds and a statement that a ton weighs two thousand pounds. There is no way of knowing whether the statement about Mr. Smith is true unless we observe the results of some measure. But it is entirely useless to observe the number to which the marker on the scale points when we weigh a ton of anything, because if it does not point exactly to two thousand then, by definition, what we are weighing is not a ton.

No factual knowledge is transmitted by analytic propositions. If I say, "Either it is raining or it is not," I have uttered a logical truth, indeed a truism, but I do not know if I should carry an umbrella because, although I have made an exhaustive statement of the possibilities of weather at any time whatsoever, I have made no factual statement about the weather at any particular time. Yet the proposition is necessarily true, or certain. It does not give information about the weather at any time, but is compatible with any factual statement about the weather, and it asserts of a pair of contradictory propositions, "It is raining" and "It is not raining," that one or the other is true at any time. Thus the whole proposition is only an application of the definition of contradiction, which states that of two contradictory propositions one or the other must be true. And to deny it would be self-contradictory.

Empirical propositions contain factual knowledge; therefore they are never necessary or certain. Their truth is only probable. If they are particular or singular propositions, they depend on observation, memory, and so forth, which are fallible; if they are general or universal propositions, negative instances are always possible. When I say, "It is raining," I have made an empirical assertion; it yields information about the weather at this time. But it may be false; at best it is highly probable. Here we should point out that even though men may want knowledge to be certain before they risk acting on it, they have always taken risks on the basis of probability. The higher the degree of probability, the less the risk, but risk of some sort there always is. The best security that can be offered man is a highly probable scientific statement.

The demand for "knowledge" in the sense of certainty has led to two extreme views: that there is a "higher" truth than science can offer, usually arrived at by intuition or by logical deduction from self-evident axioms; and skepticism as to the possibility of any knowledge whatsoever. As for intuition, the trouble is that different men have different and opposed intuitions. If we want to know which one to accept, we shall have to wait for an intuition of our own, and that carries psychological conviction, but only to ourselves. Besides, conviction is no substitute for evidence; every folly convinces some people. So we shall have to appeal to something objective, like scientific test. And if we start with self-evident axioms, then we have only analytic statements, for that which is self-evident contains its own evidence, and can only contain it as part of its meaning. Deduction from analytic statements alone yields more analytic statements, never anything factual.

As for the skeptic, we must ask what he is skeptical about, what he denies. If he merely denies the possibility of certainly true factual knowledge, he is justified. But if he denies that there is any knowledge at all unless it is certain knowledge, it may be either an expression of his unwillingness to use the same word, "knowledge," for both the analytic and the empirical, which is surely not a serious matter, or it may be a refusal to act without certainty, on the ground that probable knowledge is as bad as no knowledge at all.

What we call probable knowledge, the skeptic argues, is based on past observation, and there is no reason to believe that what held in the past will hold in the future. Actually, this is true, and even if we assert that nature is sufficiently uniform for us to expect recurrences, it can be answered that that, too, was true in the past and may not be in the future. Yet this theoretical attack on science has little to do with belief as it is expressed through action. The skeptic may reject scientific conclusions during his office hours (or, as David Hume put it, when he is in his chamber) and never be refuted, yet when he leaves his office on the fifth floor, he does not walk out the window but uses the stairs or the elevator, thus acting on his belief in the law of gravitation. Without what Santayana calls "animal faith" a man cannot survive at all.

Some skeptics have reached the point of solipsism (*solus,* "alone" + *ipse,* "self"), a belief that one can know only that he himself and his sensations exist. Of course, such skeptics have to act as though they believed in the existence of others and of the natural world. Otherwise they would be in an even worse fix than the two professors, both solipsists, who passed each other on campus and said "Good morning" in a shamefaced way because neither believed that the other existed.

Whatever the logical difficulties in understanding induction, the solipsist does not wrestle with them. Like everyone else, he lives in a world he does not know with certainty, and he merely redefines all the things in that world as the objects of *his* experience, only his own sensations. So the solipsists who greet one another might say they are greeting their sensation of another person and receiving the sensation of a greeting in reply. But with this qualification about what he perceives, the solipsist behaves like anyone else, coming in out of the rain to escape the sensation of being wet and studying science to learn about his sensations of the world.

4

# Empirical Methods of Science

The empirical methods of science are discovered by examination of the most successful scientific work, and they are stated in a general form so that they can be applied to further scientific work of all kinds. The generalization of empirical methods is the logic of scientific procedure; it states, among other things, the logical grounds for scientific conclusions. This logic or methodology of science is to be distinguished from the psychology of the intellectual process which states the psychological conditions for discovery.

There is a popular myth that great ideas, resolutions, and discoveries come about by contemplation of the homely things of life. Archimedes took a bath and noticed that his body displaced water; he then ran naked through the streets shouting, "Eureka!" ("I have found [it]"). Newton watched an apple fall from a tree and recognized the work of gravitation. James Watt saw the lid of a teakettle move when the water boiled and learned there is power in steam. The stories are endless and are told to children with the idea, perhaps, that if they learn to look at things with care and keep their wits about them, they too will discover some great matter. But Newton did not generalize from the fall of an apple to the law of gravitation; he discovered gravitation even in an apple's fall.

Great thoughts necessarily occur, like birth and death, amid the ordinary furniture of life, and to assume that an idea that comes *after* drinking tea is the *result* of drinking tea is to commit the fallacy of *post hoc ergo propter hoc.* The persistence of the myth may be due to the delight with which some of us identify ourselves with our heroes (we, too, drink tea), and to the pleasure which others take in denigrating great men (if we cannot do it, too, it is not worth much).

But why dismiss an idea because it contains this fallacy? Is not science itself based on a form of *post hoc ergo propter hoc,* in that generalizations result from observation of instances in the past and are then used to predict events in the future? And there is no assurance that because something was true in the past it will be true in the future. How, then, does scientific inquiry differ from the simple instance of the fallacy in which savages, seeing an eclipse of the sun, beat their tom-toms until the sun shines again and then attribute its reappearance to the music of their drums? It differs in three essential ways. First, the scientific generalization is based on a great many past instances. But one might object that the savages could beat their tom-toms during every eclipse of the sun through history, yet their music would be unconnected with the end of the eclipse. So, second, the generalization is made only after an attempt to establish the *relevance* to the event of the presumably causal factor. To determine the relevance of tom-toms to the shining of the sun a scientist would keep the drums silent during an eclipse (if he had enough courage) and, since the sun would reappear, he would know that the tom-toms were irrelevant. For science, the presence of *A* before the occurrence of *B* is not enough to warrant the conclusion that *A* caused *B;* it must also be established that in the absence of *A, B* does not occur. This is the essential test that the existence of *A* is relevant to the occurrence of *B.* Third, additional support is given every scientific generalization by correlating it with other generalizations that have already been established. If it fits logically into a system, then the evidence for the other generalizations adds to the weight of the evidence for this one.

Another oversimple way of explaining scientific advance is by attributing it to genius. This prevents fuller explanation, and probably has the merit of being true. But its truth is tautological and it begs the question when it is used as explanation of discovery, for anyone who makes an important discovery is called a genius, thus implicitly defining genius. Of course, if we defined geniuses on other grounds, like their scores on I.Q. tests, and then tried to learn empirically whether people who make important discoveries are geniuses, thus defined, we would avoid logical error, but we would still learn little of consequence. If it did turn out that great scientists are geniuses, we would know only one factor in scientific discovery because there are many geniuses who do nothing scientific and others who make no discoveries. We would still have to ask what the logic of scientific procedure was in the work of great scientists.

Why does circularity in argument, or begging the question, invalidate the argument? Surely the same fallacy occurs in science itself. It does, to some extent, like *post hoc ergo propter hoc,* but only in analytic matters of definition and deduction, not in factual ones. When we define word *A* in terms of words *B, C,* and *D,* we then have to define the latter, and we may do that in terms of word *A.* Anybody who looks up a word in a small dictionary and finds it defined in words whose meanings he does not know is likely to discover that, when he looks up the latter, they are defined in terms of the original word. Dictionaries are inevitably circular, but large dictionaries interpose a lot of words before returning to the first one: *A* is defined by *B, B* by *C, C* by *D, D* by *E, E* by *F,* and only then is *F* defined by *A.* What is the advantage of extending the circularity in this fashion? It is that as soon as we know the meaning of any one of the intermediate words in terms of its referent in experience, we have broken the circle. The circularity of definitions is ended only by pointing to an object which one of the words means, and although a dictionary does not literally point to any objects, once we make the mental connection between a word and the object to which it refers, we have, intellectually, broken the circle.

But circularity in definition, or in deduction, cannot be escaped so simply in a system of formal logic or mathematics. Logical and mathematical symbols do not refer to things in experience and, if we enter a logical circle, we cannot break it by mentally pointing to a meaning outside the circle. And we establish the validity of a statement only by using another statement whose validity in turn is established by still another. There would seem to be only two possibilities. Either we continue the process in an infinite regression, which would never finally establish anything, or we establish the validity of some statement by using a second statement whose own validity has already been established by the first statement, thus arguing in a circle.

In constructing a formal science, therefore, we begin with a small number of *undefined terms* that seem immediately understandable, perhaps because of their familiarity from other uses. These are called *primitive terms.* We then define other terms by the use of primitive terms. This is like the way we break the circle of dictionary definitions, but it has another element. As we use the primitive terms in a number of formal statements, their meaning is clarified, as the meaning of words would be when used in sentences, by their context.

Similarly, in the *statements* of a formal science, we start with *primitive statements* or *axioms.* These are simply to be accepted, although their validity has not been established. And we know what they mean because their terms have been defined or are primitive. Then we proceed deductively, accepting no other statement until it has been proved on the basis of axioms and already established statements. Thus there is neither circularity nor regression. Of course, nothing in the system has been proved to be true, actually to give knowledge of the world, so long as the axioms have not been proved to be true. But that is the nature of a formal science and that is why its statements are at best valid, not true.

So we can dismiss explanations of scientific discovery, like genius, which are circular. Yet we cannot give a satisfactory explanation, because the causes of discovery have not been thoroughly explored. What is clear is that many discoveries are based on accidents whose meaning is grasped by a scientist psychologically prepared to do so. The logical pattern of investigation is then obscured by luck and insight, but it underlies them. The frequency of such accidents in science may be gauged by the fact that a special word, "serendipity," has been adopted; this is defined as the gift of finding unforeseen things of value. The history of the United States is in part a result of Columbus' serendipity.

An outstanding scientific instance of serendipity is associated with the anatomist Galvani, whose name gave birth to the word "galvanize." Several frogs' legs, to be used in one of his experiments, were hanging from an iron balustrade in Galvani's home. They were swung by the wind and from time to time touched the balustrade; each time they touched, they twitched. Investigation of this curious phenomenon revealed that the legs were suspended by a copper wire and the contact of copper and iron produced an electric current. A contemporary of Galvani named Volta (from whose name come "volt" and "voltage") was then led to experiment with dissimilar metals to produce electric currents, and so invented the electric battery. Two other examples of serendipity may be mentioned briefly. Pasteur discovered his method of immunization by accidentally inoculating fowls with an old bacterial culture that had little virulence. Alexander Fleming noticed that a culture of pus-producing bacteria which was accidentally contaminated by a mold started to dissolve. Following this hint, he discovered the curative functions of penicillin.

Above all, instead of using serendipity as evidence that there is no real method in science, we can use it as evidence that discoveries, no matter how accidental their bases, are made by men who know what to do to follow up a clue. This implies that there is scientific method, and that training in it is a requirement for accomplishment. Serendipity has its place in the psychology of discovery, but it occurs only when a lucky accident is fitted into the logic of discovery.

Examination of the empirical—or hypothetico-deductive—methods of science yields a general pattern of scientific inquiry, which can be described as a progression through five stages. They are:

*First,* statement of the problem;
*Second,* formulation of a hypothesis;
*Third,* deductive elaboration of the hypothesis;
*Fourth,* test of the hypothesis;
*Fifth,* acceptance, rejection, or modification of the hypothesis.

The first stage, the statement of the problem, is one for which there are no precise rules. We start with awareness of some difficulty. Difficulties do not come to us in experience as neatly stated problems; we must articulate them so that they are questions clearly asked and capable of solution.

Difficulties are felt and resolved; problems are stated and solved. We resolve a difficulty by action, but we do not know how to act until we have formulated the difficulty as a problem and then have reached a solution. If I am sick and want to be well, I must first ask what is the matter with me, and having learned the answer, I must ask how it is cured. When these problems are solved, I can act effectively. Within the scientific context, purely intellectual difficulties are more usual. We become aware of a logical contradiction, or we feel that an accepted belief does not work as it should in practice, or we find vagueness or ambiguity in an idea. Or we wonder why no difficulty is felt—why, for example, I feel well when I am healthy—and this concern itself may be thought of as a difficulty to be converted into a problem. We will not get very far if we find difficulties only when something is wrong. If I do not understand the conditions of health, I will have less chance of knowing what is wrong when I am sick; if I do not know why the economy is prosperous, I am less likely to understand why, at another time, it is depressed. However, whether we begin with health or sickness, full understanding of one requires understanding of the other.

It is relatively easy to know that there is a difficulty, even to have a vague idea where it is, but it is sometimes a great intellectual feat to state it as a problem. And years can be spent working on the wrong problem. How can a problem be wrong? After all, a problem is a question, not a proposition, and so cannot be either true or false. But it can be wrong in the sense that it does not formulate the underlying difficulty—or formulates it inadequately. So even if it is solved it provides no guide to action which would resolve that difficulty. On the other hand, some difficulties that bedevil science for years are readily eliminated once the problem is stated correctly.

A significant advance in chemistry was made when Antoine Lavoisier discovered the nature of combustion, partly as a result of correctly stating the problem. Formerly chemists had believed that combustion required an element called "phlogiston" which was contained, supposedly, in all inflammable materials; the liberation of phlogiston was regarded as the process of "burning." It was Lavoisier who realized the inadequacy of this solution because it in no way removed the initial difficulty: there is always an increase in weight as a result of combustion. The theory that phlogiston was liberated implied that there was a decrease in weight.[1] The problem that other chemists had tried to solve was: What is the nature of combustion? Stated thus, the problem is so vague that no sort of hypothesis is suggested. What Lavoisier did can be regarded as a restatement of the problem as a result of a better understanding of the difficulties from which it arose. He asked what it is that is added in combustion, since

[1] Some chemists of the time provided a fine instance of a theory that was beyond test, thus violating the verifiability criterion of meaning. They argued that since burning was the liberation of phlogiston, phlogiston must have negative weight, and the more that was liberated, the greater the increase in weight. The theory of negative weight, used in this way, is also an example of the fallacy of *petitio principii*, or begging the question.

there is more weight after combustion than before.[2] In this form the problem suggests certain hypotheses, such as that oxygen, in the absence of which burning does not take place, combines with the combustible substance. And that is what Lavoisier discovered to be true.

The second stage in scientific inquiry, the formulation of a hypothesis, depends on the problem that is stated. A hypothesis is a proposed answer, or solution, to a problem; it is an informed guess or idea. The two basic rules of a good hypothesis are that it answer the problem and that it be testable, directly or indirectly. A well-formulated hypothesis leads the scientist to just those facts which are relevant, that is, which test the hypothesis. An erroneous stereotype of science, on which we have already commented, is that when a scientist has a problem he starts collecting "facts." Actually, from a problem alone one does not know what "facts" to look for. The world is full of "facts" and it is useless to go about collecting them.

The hypothesis leads the scientist to relevant data by way of the third stage in the pattern of scientific inquiry, the deductive elaboration of the hypothesis. Unless the hypothesis is in the form of a particular statement about observation, it cannot itself be tested directly. We can deduce from it, however, such particular statements. If these are tested and found true, they are evidence for the truth of the hypothesis.

We may take an example from social science. Experience with Negroes in the North and with Negroes in the South led to the question: What are the differences between the two groups? One of the things discovered is that Northern Negroes appear more intelligent on the average than Southern Negroes even when all the Northern Negroes tested were born in the South. As with so many answers to questions in science, this one led to a stated problem: What is responsible for the difference in intelligence of those groups? One plausible answer (hypothesis) is selective migration; the brighter, hardier, more ambitious Negroes leave the South to take advantage of greater opportunities in the North. Another plausible hypothesis is that the better

[2] It may seem to the layman that there must be a decrease in weight when something is burned. Can a heap of ashes weigh as much as a man? But the ashes are not all that is left after combustion; there are also the gases that normally escape. When everything is weighed, there is an increase in weight.

conditions under which Negroes live in the North permit a fuller use or training of their intelligence.

Neither of these hypotheses can be tested directly; they are generalizations, not particular propositions. What can we deduce from them that can be tested? And what test will allow us to choose between the two? If selective migration is the correct hypothesis, then it follows that Negroes who have just come to a Northern state have the same intelligence, on the average, as Negroes who have lived there for years. If environment is the correct hypothesis, we can deduce just the opposite; namely, that old residents will have a higher I.Q., on the average, than recent arrivals. (This, in fact, turns out to be the case.) Our deductions, then, have led us to the relevant facts—the average I.Q. of recently arrived Negroes and the average I.Q. of old residents. And we learn, when we compare them, which finding does confirm which hypothesis.

The fourth stage, testing, always rests on observation. The objectivity of science—or its "intersubjectivity" if one prefers that word—depends on its being testable by anyone. With sufficient training, everybody who makes the same inquiry and repeats the same experiments should come to the same conclusions. The observations on which the test rests should be the same for all people with normal senses.[3]

And what should be observed? What is relevant. The criterion for relevance may be stated this way: An observation is relevant to an inquiry when the description of what is observed is a proposition which is evidence for or against the truth of the hypothesis being investigated.

Sometimes the observation required to bring about the test of a hypothesis is easy enough to make, but often a great many conditions must be exactly controlled. At times the only observation that tests the hypothesis is observation of the results of experiment, that is, of a deliberate manipulation of the materials of the inquiry for the purposes of test. Experimentation is a subject in itself, much of which goes beyond our purpose, but a few general comments should be made. Science is plagued by *ceteris paribus,* "other things being

[3] For a short, precise description of verification see Rudolf Carnap, *Philosophy and Logical Syntax,* Kegan Paul, Trench, Trubner and Co., Ltd., London, 1935, pp. 10-13.

equal," which is assumed even in the most seemingly unqualified predictions. There will be a total eclipse of the sun at such and such a time, *unless* planets collide, the sun explodes, or there is some other catastrophe. Under current conditions, there will be a movement of people from country to city, *unless* industry is decentralized, the government intervenes, farm prices go up, there is a war, and so on. And, since all other things are seldom unchanged, our social predictions are hazardous, like the situation of a man crossing the street in heavy traffic who can see only one car at a time. *Ceteris paribus* can be established, not just assumed, by experimentation, for experimentation at its best controls the variables it needs and rids itself of variables that might interfere.

When one deliberately manipulates materials in order to see what results, he must minimize the possibility that the same results would have occurred had he let things take their natural course. The usual way of accomplishing this is by the creation of a control group or situation, that is, a contrasting group or situation with which the experimenter does not interfere. A decisive use of experimental control is found in an inquiry by Francisco Redi, results of which were published in 1668.[4] The problem at issue came from the common observation that when meat was exposed for several days, worms appeared in it. Some people regarded this as an example of spontaneous generation. Redi believed that the "worms" hatched from eggs laid by flies. Observing again, and more carefully, what had led to the problem, he saw small black flies hovering about the decaying flesh, and not only worms but eggs on its surface. And he proceeded to test his hypothesis experimentally by eliminating the flies and observing the meat over a period of days.

Redi's procedure was simple: he sealed some meat in glass flasks *and left other meat in similar flasks that were unsealed.* He then observed that, though the meat in the open flasks "became wormy and flies were seen," no worms were visible on the meat in the closed flasks. Thus he eliminated the possibility that worms would not have been produced at this time even if he had not sealed his flasks. Still another possibility existed, though. No air circulated in the closed

[4] This and other important experiments in science are treated simply and well in James Bryant Conant, *Science and Common Sense,* Yale University Press, 1951.

flasks. Perhaps this, rather than the absence of flies, kept worms from the meat. So Redi tried again, this time not sealing the flasks but covering them with a "fine Naples veil that allowed the air to enter," but kept flies off. Result: no worms.

Now we are at the final stage in the pattern of inquiry. If a hypothesis has been confirmed by test, we accept it as true, but as probable rather than as certainly true, and it remains subject to further test. If a hypothesis is disconfirmed by test, we reject it and try to formulate another and more adequate one. But sometimes the results of the test are not quite what we predicted and yet are close to it. This usually leads to a modification of the original hypothesis. Then we follow the pattern once more with another set of deductions and new tests.

Let us examine a social problem, using, as we must in any complete inquiry, both analytic and empirical methods. Suppose we ask the question: "Why is India poor?" Exactly what do we mean by this question? What answer would satisfy us? We may even go a step further: Why is this a question that interests us?

We begin by analyzing our problem. As stated, it presupposes that India *is* poor. Now, the appearance of poverty is relative to the appearance of wealth; India seems poor to us because it is less well off than our own country in some economic ways. And this appearance probably gave rise to the problem; something was different from what we are accustomed to, and we are likely to take the accustomed for the normal and to regard the unaccustomed as a problem. A little reflection might lead us to reverse the terms and ask: "Why is America rich?" If we discover why India is poor we may know better why America is rich, and perhaps that is why the first question interests us.

How can we clarify the meaning of "poor"? When we say India is poor we do not mean that all Indians are; some Indians are rich as Rockefeller. Nor do we mean that the Indian government is poor; it may have a surplus in its budget while ours has a deficit. So we must decide on some criteria or operations which will measure the economic welfare of a country. These probably include the amounts of goods and services at the disposal of the average inhabitant and how long and hard he has to work for them. Here we might examine the statistics (based on empirical data) on national income, divide the total national income by the total population, and translate it into real income. If we did this with both India and the United States then, with all necessary adjustments and qualifications, we would doubtless discover that the average Indian is much less well off economically than the average American.

Our problem is to find out why this is so, and there are many kinds of answers (hypotheses) which may be given: size of population, amount and quality of natural resources, political and economic organization, and so forth. One answer popular in this country for some time was that India is still poor because of British exploitation.

What do we mean by "exploitation"? Some people say they are exploited when they mean they are underpaid, and of course most people think they are underpaid. Others use "exploitation" as a synonym for "poverty," apparently assuming that poverty is always due to exploitation. Neither subjective feelings nor overgeneral assertions will help us.

There are two useful definitions of exploitation. First, we could say that a class or group is exploited when it receives lower wages or prices than it should in purely economic terms, because some other group or class has intervened in an extraeconomic way to take away part of its expected income without giving services or goods in return. Of course, it would be very difficult to calculate just what the first class or group should receive when there are no actual instances of people receiving it. The second definition avoids this difficulty. We may say that a man is exploited when he gets lower wages or prices than are current for the situation he is in. A worker, then, is exploited when his wages are lower than those of other workers who do the same sort of job, a farmer is exploited when the prices for his produce are less than the going rate, an author when his royalties are less than those of equally known and able writers for the same sort of book.

According to either of these uses of the word "exploitation," people are exploited only when they are prevented from obtaining current or properly expectable wages, prices, or royalties; and exploitation is unlikely unless a person or group benefits from the exploitation and has some sort of power over those exploited. This is not the same as someone making a profit in the course of ordinary business procedure; it involves special conditions which enable the exploiter to benefit beyond any profit derived in the ordinary way.

Now we have formulated and tried to understand our problem and the kind of answer that would satisfy us. We have proposed a hypothesis—or rather, mentioned one that had great currency. We have analyzed that hypothesis to discover what it means, and are now prepared to elaborate it deductively so that it will lead to evidence for or against its truth. What follows from the hypothesis about the economic relations between England and India that we can test? If England exploited India, then the prices England paid for Indian goods were lower than those current in world markets not under English domination, the prices England charged India for British goods were higher than those paid by others, and the wages paid Indian workmen by English management were lower than the wages paid Indian workmen under similar conditions by management of other nationalities.

Can we find evidence that will test these deductions? If Indian goods were sold only to England and British goods only to India, and if Indian workers were employed only by Englishmen, it would be very difficult to make a satisfactory test. We could use the definition of exploitation as paying less than purely economic considerations would dictate, and we would then have to find a way of discovering what would have been paid as a result of those considerations alone. But in fact British goods were sold everywhere and many, if not all, Indian goods were sold throughout the world.[5] In addition, not all businesses in India had English management. Not only were nationals of other countries represented, but the employer class was largely Indian. So we can compare the prices England paid and charged, and the wages they paid, with those current on other markets and under other management (empirical evidence). If it turns out that the English paid less and charged more, we would have to show further that the Indians did not on the whole buy from and sell to others because England prevented them from doing so. Having shown this, we could conclude that England exploited India; but if the Indians maintained their economic relations with England of their own volition, we could not say they were exploited, but only that their economic behavior was foolish.

Now we know what evidence would test our hypothesis and we can go on to examine the evidence. When we do so we discover that the prices and wages charged and paid by England were not substantially different from those current elsewhere. England probably did

[5] Or else the same, or similar, goods, though produced elsewhere, were sold on world markets.

have the power to exploit India in the fashion we have described, but she apparently did not use it. So we have found the hypothesis to be false, and in order to answer our problem we must propose another hypothesis.

If it be objected that our definition of exploitation is too narrow, that there are other forms of exploitation besides this special economic one, and that England probably did exploit India in other ways, we can answer only that to have used a definition less limiting would have made it impossible to get a clear answer. If one wants to talk of political and social exploitation as well as economic, then each of these would raise another problem to be tested separately. Perhaps England failed to help India as much as was in her power or prevented India from helping herself. Perhaps the political development of India was retarded by British rule, her culture impaired, her nationals enslaved. Each of these is a problem in itself.

But if other economic factors are introduced, do they invalidate our conclusion? Perhaps Indians paid high taxes to an English government that did little for them in return. Now that India is autonomous, we can compare current taxes and government services with those under British rule. It is unlikely that we will come to a definite result. But we need not pursue this subhypothesis, because the whole Indian government budget was surely too small a part of Indian national income to have been the cause of poverty. Other things may be urged. England may have robbed India, that is, taken things without compensation. Even though there were instances of this sort, they were economically insignificant. The hypothesis that India's poverty was due to English exploitation means not that there was some exploitation here or there but that there was enough to influence the Indian standard of living significantly.

So our conclusion stands: the hypothesis is false. In offering another hypothesis we may benefit from the elimination of this one. Perhaps we have learned in the course of examining the evidence that exploitation of some kinds is insufficient as a cause of poverty, or that it is a symptom of a more basic organization of production and distribution which is inadequate for prosperity. If so, our new hypothesis will take this knowledge into account. The hypothesis we examined about India's poverty is, in fact, only one of many that might be proposed. But if there were only two possible hypotheses, each contradicting the other, the elimination of one would be evidence that the other was true, because of the law of logic that whenever there are two contradictory propositions one must be true and the other false. Thus, of the following two hypotheses, "All people are afraid of snakes" and "Some people are not afraid of snakes," if either is shown to be false the other must be true. "Some people," remember, is interpreted as "at least one person"; and if it is false that there is at least one person not afraid of snakes it must be true that all people are afraid of snakes, just as if it is false that all people are afraid of snakes, there must be at least one person who is not afraid of them.

General hypotheses assert something about all members of a class. All highly confirmed general hypotheses are scientific laws, and all scientific laws are highly confirmed general hypotheses. The establishment of laws is the aim of science because, as we shall see, without them we can neither explain nor predict.

A simple example of scientific law is what Archimedes actually established in his famous bath; if a solid denser than water is immersed in water, it will lose weight equal to the weight of the displaced water. Another example is the Law of Falling Bodies established by Galileo through dropping weights from the Leaning Tower of Pisa: All bodies fall at the same rate (in a vacuum) and at the end of any specified time have a velocity proportional to that time and have traversed a distance proportional to the square of that time.

Note that Archimedes' law does not state that any solids are, in fact, ever immersed in water, nor does Galileo's law state that any bodies ever fall in a vacuum. They are both hypothetical, like all scientific laws; they mean, respectively, that *if* any solid denser than water is immersed in water, or *if* any bodies fall in a vacuum, then what the law states will be true of them.

A scientific law asserts that whenever there are specified *kinds* of conditions, a specified *kind* of event will occur. On examination the conditions turn out to be of two types: necessary and sufficient. The traditional definition of the *cause* of an event is the necessary *and* sufficient conditions for its occurrence.

A necessary condition for an event is a condition in the absence of which it will *not* occur; a sufficient condition for an event is a condition in the presence of which it *will* occur. Necessary conditions alone never bring anything about. It is necessary to have eyes and a minimal intelligence to read this book, but in addition one must open it and go through it. Oxygen is a necessary condition for the occurrence of combustion: there is no combustion unless oxygen is present. But the mere presence of oxygen is not sufficient for the occurrence of combustion, or there would be fires everywhere. Ignition, friction, spontaneous combustion may all be sufficient conditions for the occurrence of combustion in that, granted the presence of oxygen, the presence of any *one* of these is sufficient for combustion to occur. If we are trying to ascertain the cause of a particular fire it will be enough to discover what started it, that is, which of the possible sufficient conditions was present. We need not add the existence of necessary conditions, like oxygen, because there *was* a fire, so the necessary conditions must have been present. But if we are trying to state a general law about combustion, neither ignition nor friction nor spontaneous combustion can be called *the* sufficient condition of combustion in general. They are only particular ways in which that condition appears. So we have to find out what they have in common, perhaps a specified increase in the rate of molecular motion.

The difficulties in finding a sufficient condition general enough to be part of a law when there is a multiplicity of particular sufficient conditions can be seen in the Hawthorne experiment, started in April, 1927, at the Hawthorne Works of the Western Electric Company.[6] An earlier study of the effects of illumination on the productivity of workers had had an unexpected result. An experimental room of workers and a control room had been set up; lighting was first increased and then diminished in the experimental room, while it remained constant in the control room. When lighting was increased, production rose. Simultaneously it rose in the control room. When lighting was diminished, production also rose in both rooms.

This seemed to make no sense, so a test room was set up again and maintained through a larger number of experimental changes. First the workers were invited to conferences about each proposed change. Then the conditions of work were altered, one at a time, for twelve periods of varying length: more rest periods, longer rest periods, shorter work day, shorter work week, food during morning rest periods, and more pay. The twelfth period was a return to the original conditions of work. In every period the daily and weekly output rose, reaching its height in the twelfth period. The workers then returned for

[6] F. J. Roethlisberger, *Management and Morale*, Harvard U. Press, 1942.

thirty-one weeks to the conditions of the seventh period, which included a fifteen-minute rest with food in the morning and a ten-minute rest in the afternoon. Output reached a new high.

Every condition introduced was sufficient for greater productivity, even a return to original conditions. What did the conditions have in common that could be stated in a general law? All they had in common was that they were parts of an experiment about which the workers knew and had been consulted. Simply because there was an experiment the workers felt that their employers were interested in them as human beings, and this was sufficient to bring about increased output.

Confusion between sufficient and necessary conditions is common, and so is a mistaken identification of cause with either necessary or sufficient conditions, whereas both are needed. If, for example, we say that the family came into being because of man's sexual drives, we have substituted a necessary condition for a cause.

The existence of sexual drives is a necessary condition for the existence of the family; if there were no sexual drives the institution of the family would not exist. But obviously the satisfaction of sexual drives, or their mere presence, does not require that the family exist, for sexual drives may be repressed, or satisfied outside the family.

Requests that something be explained may mean "Make this more simple," "Restate this," "Tell me how this action was performed," "Why did you do it?" and so on. Literally, none of these asks for scientific explanation; they ask, respectively, for simplification, paraphrase, description, and motive. A request for scientific explanation asks about an event or phenomenon. "Why is there an eclipse of the sun?" "Why did Josiah Terwilliger spend devalued currency but save hard money?" "Why do Japanese mothers put up with willful behavior by young sons?"

The basic principle of scientific explanation is that events are explained when they are shown to be instances of a general law, or, more technically, when a statement of them can be deduced from other statements which refer to a general law. So we must show two things. First, there must be a scientific law (or laws) which states that whenever events or conditions of a specified kind exist, events of the kind to be explained will take place. Second, particular conditions of the kind described in the law must actually have been present.

These two statements comprise an explanation. We can deduce from them a statement that the event to be explained took place—in other words, that the event was an instance of a general law.

Thus if we want to explain why the birth rate in the United States rose during the years 1941-45, we can say, "Whenever nations are at war their birth rates rise, and the United States was at war from 1941 to 1945." Assuming the generalization to be true, this statement is an explanation of the rise in the birth rate because we can deduce that rise from it. Still, someone may say, "This is no explanation at all; it is only a description of what happens. What we want to know is *why* birth rates rise during war." This criticism does not invalidate the explanation; it only asks for still another explanation, this time an explanation of a general law rather than of a particular event. If we know enough, we can give the additional explanation, but it will follow the same pattern. We will show that the law itself is an instance of a still more general law. For example, if we know that tensions of a specified kind yield a propensity to have children and that those tensions always exist when there is war, we have offered an explanation. And we can go on in this fashion to explain each law as an instance of a still more general law, to the limits of human knowledge.

Explanation and prediction are opposite sides of the same coin. The questions are different, of course, but the complete answer to either of them will be based on the same pattern. When we ask why an event occurred, we want an explanation. If, however, we do not know that the event occurred but do know the conditions of its occurrence and the existence of those conditions, we can predict the event. So if we know a law that birth rates increase during war and know further that a particular case is an instance of the conditions of the law—*this* nation is at war at *this* time—we can predict that an instance of the event that occurs under those conditions—increased births—will also occur.[7]

---

[7] In summary:

EXPLANATION

1. Event *E* to be explained

   requires

2. Law that if condition *C* occurs, then *E* occurs

   and

3. Statement that *C* in fact has occurred.

Ability to predict often gives ability to control. Science is a quest for knowledge, but men are concerned also with the uses of knowledge. And the attempt to control events raises questions that go beyond science: What ends do we wish to attain? For what purposes do we want control? What means are legitimate? We will consider these questions later; here we want only to show how science can yield control.

We learn from science that under specified conditions a particular event will occur and we may also learn how to create those conditions; thus we can bring about the event. On the other hand, if the conditions can be prevented from occurring, we *may* be able to prevent the situation from occurring. If we know that frustration brings about aggressive action of a hostile nature, an attempt to injure others, then we know we can create hostility in people by frustrating them. We may even know how best to frustrate—for example, by thwarting people in the expected attainment of goals. So we can make soldiers hostile before they go into battle by subjecting them to a period of frustration. But we know also that the first object of aggression is whatever causes the frustration, so we must be careful that the soldiers do not want to turn on their officers. To avoid mutiny, we must create still other conditions that will displace the aggression and turn it on the obvious target, the enemy.

However, if we want to eliminate race riots and tensions in a city that has a history of both, we must first know that these acts of aggression did in fact come from frustration, because we may not be sure that *only* frustration leads to aggression. If we decide that the riots are a result of frustrations, we must find out what those frustrations are and see if they can be eliminated. Yet once we know the cause of the riots, we may conceivably change our minds about stopping them. Our decision will depend on the value we place on race harmony as opposed to the value we place on maintaining the conditions that caused race disorder. Those conditions may have had other consequences which we think desirable, and elimi-

---

PREDICTION

1. Law that if condition $C$ occurs, then $E$ occurs
   and
2. Statement that $C$ occurs leads to prediction that
3. $E$ will occur.

nating the conditions may not only eliminate race disorder but the desirable consequences as well. How we use instruments of social control is a matter of values and morals, and the better the instruments the greater their potential for both good and evil. But social control itself depends on knowledge, essentially knowledge yielded by social science, and it is to social science and its difficulties that we turn now.

5

# The Scientific Study of Society

Can there ever be a science of society as there is a science of nature? Can scientific method, or any variation of it as it now exists, be applied to the study of human behavior? There are many who think not.

The major objections to the possibility of social science resolve themselves into the following: (1) much social science, especially sociology and social psychology, merely elaborates the obvious; (2) men are irrational and their behavior cannot be accounted for; (3) much human behavior is private, hence not directly observable, and experimentation is often impossible; (4) the social scientist, like all men, adopts the values of his society; this prevents him from studying society objectively; (5) social behavior is so complex that we cannot, as we can in physical science, isolate the relevant variables; (6) social events, unlike physical occurrences, are unique, and science is impossible unless it can deal with the recurrence of events and experiment; (7) social predictions influence behavior and that influence may bring about what is predicted—or its opposite—thus making it impossible to know whether the prediction was initially correct. In principle, therefore, social predictions cannot be as reliable as physical predictions.

As arguments that social science is impossible, these objections can all be met; as difficulties which must be overcome if there is to be social science, they are very grave indeed. The difficulties are the reasons that social science is perhaps no more advanced than natural science was in the Hellenistic era.

The first objection, that social science elaborates the obvious, is commonly made by sophisticates. They find it trivial to read in works of sociology and social psychology that the rich have different attitudes from the poor, that men are influenced by their families, that people become angry when frustrated. Why the expenditure of time and money to learn these things, they want to know, when no one has ever doubted them? Now, it is true that such pronouncements are made in serious studies and that everybody but the writer seems to know the conclusions in advance. But there are two justifications for the statement of what seems obvious.

The obvious sometimes has to be stated as a basis for what comes next. To say that the rich have different attitudes from the poor may be just a preliminary to detailing the differences, some of which are by no means obvious. Some experiments in perception—challenged by other investigators—suggest the startling conclusion that coins actually look larger to poor children than to rich ones. And sometimes the full meaning of the obvious escapes us unless it is stated explicitly. Also, what seems obvious is often not obvious at all.[1] What is discovered by sociological and psychological study is often called obvious after the discovery has been made, when it would not be guessed in advance.

To make this general point, Professor Paul Lazarsfeld compiled a list of statements typical of discoveries about the American soldier made by the Research Branch of the War Department's Information and Education Division. He assumed that each one would seem obvious to the layman, and he added in parentheses the layman's probable justification: [2]

[1] Even that great arbiter of belief, "common sense," changes from age to age and includes the scientific conclusions of a bygone day. Today it is common sense to know that the earth is round, just as once it was common sense to know that it was flat. Many of our most fondly held beliefs about society are probably like the notion that the stars are the eternal fire that rings the universe seen through rifts in the blue cloth of the heavens, beliefs, that is, accepted as true not because of evidence but because of repetition.

[2] "The American Soldier—An Expository Review," *Public Opinion Quarterly,* Vol. 13, No. 3, Fall 1949, p. 380.

1. Better educated men showed more psychoneurotic symptoms than those with less education. (The mental instability of the intellectual as compared to the more impassive psychology of the-man-in-the-street has often been commented on.)

2. Men from rural backgrounds were usually in better spirits during their Army life than soldiers from city backgrounds. (After all, they are more accustomed to hardships.)

3. Southern soldiers were better able to stand the climate in the hot South Sea Islands than Northern soldiers. (Of course, Southerners are more accustomed to hot weather.)

4. White privates were more eager to become noncoms than Negroes. (The lack of ambition among Negroes is almost proverbial.)

5. Southern Negroes preferred Southern to Northern white officers. (Isn't it well known that Southern whites have a more fatherly attitude toward their "darkies"?)

6. As long as the fighting continued, men were more eager to be returned to the States than they were after the German surrender. (You cannot blame people for not wanting to be killed.)

There is only one thing wrong with these responses to the "obvious." *Every one of Lazarsfeld's statements is deliberately the direct opposite of what was found.* But if the true findings had been given, the layman would probably have found them equally obvious and would have had equally persuasive explanations.

It is just as persuasive to say that "obviously" (1) better-educated men can adapt more readily to new conditions; (4) Negroes are more eager to become noncoms than whites because they need the additional prestige; and (6) men were more eager to return home when the fighting was over because the job was done and they saw no reason to stay abroad.

We can dismiss the second objection briefly. Human irrationality can be dealt with by rational methods. It is not irrationality but utter unpredictability of behavior that would make a science of man impossible. Irrationality is exhibited in behavior that is not based on logic and reason; unpredictability is exhibited in behavior which can never be foretold on the basis of any knowledge whatsoever. A man may be irrational in a way that is entirely predictable. Indeed, even the irrationality on which an action is based may often be predicted. Because science is rational, it does not follow that its human subjects must behave rationally.

Third, it is true that scientific observation and experiment often require some control of the subject matter and of the conditions of the inquiry. Such control is sometimes beyond the power of the social scientist. Observation is limited by the demand for privacy, and questions about private matters are not always answered truthfully. Experimentation is limited because it would often be inhumane, because failure might mean injury to life or personality, and because it might require too great a sacrifice on the part of the people experimented upon. How many mothers would give up their babies to psychological experiment for the sake of science? How can we dam the flow of war and crisis in order to have time and opportunity for research?

These are serious difficulties, but there are ways of minimizing them. Much that we *can* observe is valuable and some experiment *is* possible. Interview and questionnaire techniques can be so constructed that lies and equivocation are revealed. Many experiments that would violate privacy or interfere unwarrantably with society can be duplicated without harm to anyone under laboratory conditions. If we want to learn, for example, whether people tend to remember what is favorable to their belief and forget what is unfavorable, we need not follow them about, prying into intimate affairs and listening through keyholes. We can question groups to discover their beliefs about some particular subject, present them with written materials of comparable difficulty favorable and unfavorable to that belief, then test and retest them over a period of time to see what is remembered and what forgotten.

Yet here there is a real limitation of social science which can be met only in part by scientific ingenuity. We must be cautious about assuming that people behave in a laboratory situation as they do under the stresses of ordinary life. The laboratory isolates what we want to test from other things in life that are *logically* irrelevant; but those same things are sometimes not *psychologically* irrelevant outside the laboratory. And the very fact of participating in the experiment may make the subjects behave quite differently than they would in ordinary life.

Allport and Postman comment on this problem as it was raised by an important experiment in the psychology of rumor: [3]

[3] Gordon W. Allport and Leo J. Postman, "The Basic Psychology of Rumor," *Transactions of the New York Academy of Sciences, Series II,* Vol. 8, 1945, pp. 61 ff.

Our method is simple. A slide is thrown upon a screen. Ordinarily, a semidramatic picture is used containing a large number of related details. Six or seven subjects, who have not seen the picture, wait in an adjacent room. One of them enters and takes a position where he cannot see the screen. Someone in the audience (or the experimenter) describes the picture, giving about twenty details in the account. A second subject enters the room and stands beside the first subject who proceeds to tell him all he can about the picture. (All subjects are under instruction to report as "accurately as possible what you have heard.") The first subject then takes his seat, and a third enters to hear the story from the second subject. Each succeeding subject hears and repeats the story in the same way. Thus, the audience is able to watch the deterioration of the rumor by comparing the successive versions with the stimulus-picture which remains on the screen throughout the experiment. . . .

It is necessary to admit that in five respects this experimental situation fails to reproduce accurately the conditions of rumor-spreading in everyday life. (1) The effect of an audience is considerable, tending to create caution and to shorten the report. Without an audience subjects gave on the average twice as many details as with an audience. (2) The effect of the instructions is to maximize accuracy and induce caution. In ordinary rumor-spreading, there is no critical experimenter on hand to see whether the tale is rightly repeated. (3) There is no opportunity for subjects to ask questions of his informer. In ordinary rumor-spreading, the listener can chat with his informer and, if he wishes, cross-examine him. (4) The lapse of time between hearing and telling in the experimental situation is very slight. In ordinary rumor-spreading, it is much greater. (5) Most important of all, the conditions of motivation are quite different. In the experiment, the subject is striving for *accuracy*. His own fears, hates, wishes are not likely to be aroused under the experimental conditions. In short, he is not the spontaneous rumor-agent that he is in ordinary life. His stake in spreading the experimental rumor is neither personal nor deeply motivated.

When a scientist is aware of the difference between a laboratory situation and the same type of situation outside the laboratory, he must try to interpret his experimental findings, as Allport and Postman did, with an eye to those differences. But we cannot be sure that we have noted all the important differences, and it is difficult to interpret our findings in the light of those we have noted.

Fourth, it is argued that the social scientist, in adopting the values of his own society, cannot view any society objectively. In contrast, it is pointed out, the natural scientist deals with subjects to which his prejudices are irrelevant. Democrat and totalitarian, saint and sinner, Christian and atheist can agree on the law of gravitation, but they may disagree about the nature of economic depression, the reasons for social cohesion, and the social consequences of divorce.

In answer, it should be noted first that the inquirer is an element that cannot be overlooked when we evaluate the results of *any* inquiry. One reason natural science did not come into its own until the sixteenth century was that investigators of nature had all kinds of socially conditioned values and beliefs about the natural world. Some thought it was organized to bring about specific ends and that cause and effect did not operate in it; others thought that the movements of the spheres could be accounted for by the activity of angels. We now view nature objectively, quite apart from our own values; yet that is in part a result of the great success of centuries of work in natural science, which has slowly altered our conception of nature. But even with rigorous techniques of physical investigation and the utmost care, the observer may see what he would like to see rather than what is there to be seen—perhaps because he wants so much to have his observations accord with his theories. Or he may be so concerned with what is relevant to his interest at the moment that he neglects other things. Charles Darwin told of looking for evidence of a biological theory in an area which bore the most obvious marks of glacial erosion, yet he was so engrossed by his own problem that he noticed none of the geological formations until years later, when they were pointed out to him. An observer may even remember and forget what suits him. Darwin kept a special notebook to record evidence against his theories. He said he never forgot evidence in favor of them.

The difficulty of the social scientist in respect to objectivity is no different from that of any other inquirer, although it is probably more acute. It is met to some extent by training the student of social science in detachment and disinterestedness, and by having social scientists, often with different personal values, check each other's results in all cases, being especially careful where preconceptions or bias may exist.

The fifth objection is that social behavior can never be explained because it is too complex. If we ask in what way social behavior is more complex than the behavior of physical objects, we will probably be told that we usually know what is relevant to an inquiry in natural science, whereas it is very hard to know what is relevant in social inquiry. The objection then is clarified; it is not the complexity of social behavior that is at issue but the ability to isolate the relevant variables. It took many years for physics to eliminate the countless aspects of physical phenomena that are irrelevant to its problems and to focus on the variables that are relevant and invariant. So explanations of the behavior of hydrogen may use specific gravity and density as relevant and invariable. In comparison, it might seem that Father's excessive anger at Junior for some trivial action cannot be explained without mentioning Father's day at the office, his anxieties about his job, memories of his own behavior in childhood, and so on, until his whole life is involved.

All this appears too complex to handle, but only if the nature of explanation is forgotten. We have seen that explanation involves reference to a general law which states the type of conditions under which a type of event takes place. To explain excessive anger at a trivial occurrence, we may be able to rely on a psychological law that aggression results from frustration and is directed, under some circumstances, at a relatively helpless person or object. Then the particular frustration can be found and the outburst of anger at Junior is explained. The great difficulty is our current inability to measure such things as hostility. Here social science is still far behind natural science. But in principle the generality of a law properly omits many particular details. The theoretical problem is to isolate the relevant and generalize it properly. We will return to this in more detail in discussing prediction and forecast.

Sixth, it is contended that there can be no social science because social events are unique. No two wars are alike, no two economic depressions, no two events in the life of a man. Physical events, on the other hand, recur. The sun is in the same relation to the earth on a particular day every year, objects fall the same distance at the same speed; an observation can be tested by duplicating the conditions under which the observed phenomenon occurred. This distinction between social and physical events is, however, not nearly so sharp as it seems. The sun is in the same relation to the earth once every year but it is another year, it is not exactly the same sun (it has, for example, expended a good deal of energy in the course of that year), and it is not the same earth. What recurs in both physical and social events is a relationship between objects of a certain type. In formulating general laws we are interested in establishing invariant relations between types of events. Every event and every relation is unique, but it is also typical; every man is unique, but he is typical in so far as he is a man, and not merely Robert or Henry or Tom; every child is a unique person, but is a member of a type called "child." A social law about the relations of fathers and children leaves out the unique qualities of all to whom it refers. Its value for science consists in its description of the typical because then it applies to all fathers and all children.

This point is related to the discussion of social complexity; both have to do with the isolation of relevant matters. But the objections have a different emphasis, the earlier one resting on the complexity of the social, this on its uniqueness. And the uniqueness of social events has been of special significance in the objections to the possibility that history can be scientific. We shall discuss the point further when we deal with the nature of history. For our purpose here, it is the similarity in the two arguments that is important because the difficulty they illustrate is that of specifying in social science what is relevant to our questions. This difficulty is stated more sharply in the next objection.

The seventh objection is about social prediction. First, how reliable is social prediction? It is often said that the social sciences cannot predict with as high a degree of probability as the natural sciences. Various reasons are offered: society is more complex than nature; man is unpredictable or irrational—matters we have already considered. Of course, there is a difference in degree between the reliability of prediction in social science and prediction in natural science, but it is not based on these reasons, which would make a difference in *principle*. The difference is based on the *facts* of greater control and more advanced technique in natural science.

We may distinguish between a prediction and a forecast. A prediction states hypothetically that *if* such and such conditions exist, such and such an event will occur. A forecast states categorically that such and such an event *will* occur—for example, it will rain tomorrow—and so assumes knowledge that the conditions do or will exist. An economist can forecast the price of wheat at a given time if he knows the amount of wheat that will be on the market, the amount in demand, the money available, and so on. These are all economic considerations, and economics makes possible the prediction of different prices for wheat *if* the conditions are of one sort rather than another. Under conditions *A* the price will be *E*, while under conditions *B* the price will be *F*. But for an economist to forecast what the price of wheat will actually be, he will have to know what the economic conditions will actually be, and that will depend on noneconomic matters: on politics (who are elected and what laws they will pass), on weather (drought or excessive rain will cut down the amount of wheat produced), on international relations (war or the threat of war will increase the demand for wheat), and so on. Economic science alone is not usually sufficient for social forecast.

Physicists are no more able *in principle* to forecast a single event in a natural situation than social scientists are able to forecast an event in a social situation. When I throw a stone from the roof, just where will it land? The laws of falling bodies state the velocity with which an object will fall in a vacuum relative to the distance it moves. But the physicists will have to know much more to make a forecast in this actual situation: my strength (a biological matter), the direction and strength of the wind (a meteorological matter), the shape and weight of the stone (a geological matter), and so on. Since biology, meteorology, and geology are more advanced than politics and international relations, a natural scientist may *in fact* be able to forecast some actual events in nature better than a social scientist can forecast some actual event in society. But there is no real difference between them in principle.

People are often confused about this similarity in principle between natural and social science because the physicist can so readily isolate physical variables in his laboratory, and the economist cannot in the present state of human knowledge isolate human and social variables nearly so well—nor has he a laboratory. And when a social situation contains some economic element like cost or income, it is assumed that the situation is a totally economic one. So people expect the economist to be able to make forecasts when all he is able to do, of course, is to make predictions about the purely economic aspects of the situation. The physicist is not asked whether I should throw a stone from the roof—it may hit somebody—but only where it will land. The economist, however, *as an economist* is often asked whether there should be a particular tax, although his only function as an economist is to predict the consequences of having the tax and those of not having the tax. Sometimes, when the consequences seem desirable in advance and we impose the tax, we are not happy with the consequences when we get them. And we are likely to blame the economist because we think his statement of consequences made them seem desirable.

Finally, when a social prediction becomes known it may persuade people to act in such a way that they fulfill the prediction or prevent its fulfillment. Then it is very difficult to know whether or not the prediction would have been correct had it been breathed to no one until after the event. If I predict that you will be bankrupt within six months, you may take my statement seriously enough to conduct your business in a different manner, and so remain solvent; then it is hard to determine whether my prediction would have proved correct had you not known it. Of course, your new behavior should not be regarded as refuting my prediction. This is more obvious when the prediction is stated with some scientific care: *If* you continue to conduct your business as you have, then you will be bankrupt in six months. Put this way, the prediction has neither been fulfilled nor refuted, because you ceased to carry on your business in the manner specified. Yet you might have changed your business practices even if you had not known about my prediction. It is difficult in any particular case to determine the extent to which knowledge influences relevant behavior.

It is usually argued that natural science does not have this problem because knowledge of a prediction in natural science does not interfere with the test of its truth. When an eclipse of the sun is predicted, for example, it makes no difference how many people know about it; if the prediction is correct the event takes place. But we all grant that some physical predictions influence *human* behavior. If it were predicted that the earth would be struck by a comet in exactly one month, people who believed it might do many things they would not do otherwise: religious men might become atheists, and atheists might become religious; puritans might become libertines, and libertines turn puritan. But such acts would be irrelevant to the fulfillment of the prediction because, if it is true, the comet will strike the earth anyhow. On the other hand, if we managed to hit the comet with a thermonuclear missile that blew it to bits, we might never be sure whether, left to itself, the comet would have indeed struck the earth. We may not be able to do anything about an eclipse of the sun, but that is *only because we do not yet have enough power over natural forces,* not because there is any difference in principle between a prediction in natural science and a prediction in social science. The development of man's control of nature is rapidly putting him in the same dilemma about physical prediction that he has been in for some time about his social prediction.

Because we can change society more readily than we can nature, it is outside the laboratory, within the social arena itself, that we find the most significant self-justifying predictions. Consider the Marxist description of the future of society. Marx predicted growing discontent in the working class, a social revolution, and the establishment of socialism. He regarded these as an inevitable progression of stages which man could do nothing to fulfill or avert by deliberate action. Yet widespread preaching of Marxist doctrine causes unrest in the proletariat, creates and gives direction to radical political parties, and leads to deliberate preparation of a political revolution after which the triumphant workers—or, more accurately, the Communist Party, which claims to represent them—may create a socialist economy. How much of Marx's prediction would be fulfilled if there were no conscious movement to do so we cannot tell; how much of it, a scientist wonders, is science and how much prophecy and propaganda?

Any description must be thought of as a prediction when it is asked whether or not the description is true. Even if I make so casual a remark as "The grass on the lawn is green," it is reformulated explicitly as a prediction when I want to test its truth: "If I walk outside the house at a time when sufficient light is reflected from the grass so that colors are distinguishable and I look at the grass, I will see a green color." When we describe groups of people we are, then, predicting that if we were to examine them under specified conditions we would observe characteristics which would fulfill the prediction. And often when we "describe" groups against whom we are deeply prejudiced, we do it to justify treating them in a special way. Thus we may argue that it is wasteful to spend as much on a Negro child's education as on a white child's, because Negroes have a much higher proportion of unskilled laborers, and a lower proportion of college graduates and of successful business and professional men. This is an accurate description; Negroes don't as a rule get education equal to that of whites. Our description predicts what one will discover if he examines the evidence. And we act in such a way, in the education of the Negro, that our prediction comes true. Then, to make the matter doubly absurd, we use the accuracy of our prediction as justification for continuing that kind of education.[4]

There is no unified social science, and we have made only the barest beginning in working toward one. As we have seen, when we predict that if specified economic conditions exist then a particular economic situation will exist, we have to add the phrase "all other things being equal," which means that our statement is true only if the earth is not struck by a comet or we do not all die, or something of that sort. Even more important, our statement is true only if political, social, and psychological factors do not intervene. If there were a unified social science the latter qualification, which is the more likely, would disappear. We could make predictions about social events which would be based on laws of society as a whole, not just on laws of economics, politics, sociology, or psychology alone.

The individual social sciences differ in many ways. Each one may be said to have a different subject matter, like each natural science. And each social science deals with an aspect of society or of human behavior which can be studied fruitfully in relative independence. Just as each natural science has concepts, techniques, and instruments of its own, so does each social science. Or we may say that the individual social sciences are different approaches to, or ways of studying, the same subject. Just as theologians treat man as soul and biologists treat man as organism, so psychologists treat him as personality, economists as resources, and sociologists as a member of groups.[5] But these are only special ways of approaching the study of man—who cannot be reduced to any of them, but is understood only when all are considered.

Even without the full discussion each of the social sciences deserves, it should be valuable to name and define them in one place. They are usually listed as history, cultural anthropology, sociology, social psychology, economics, and government (or political science). There are, however, some disputes about this list. Anthro-

[4] On the self-fulfilling prophecy, see Robert K. Merton, *Social Theory and Social Structure,* The Free Press, Glencoe, Ill., 1949, pp. 184-93.

[5] This second approach will be found in Ernest van den Haag, *Education as an Industry,* Augustus M. Kelly, New York, 1956.

pology is sometimes regarded as a part of sociology. History is thought by some to be one of the humanities. And social psychology is alternately treated as an aspect of psychology or of sociology. (The problem of where to list social and political philosophy exercises some college administrators, but it is not important for our purposes.)

Briefly, history is the study of man's past; cultural anthropology, of preliterate peoples; sociology, of the behavior of groups and the relations among groups; social psychology, of the relation of men to groups and to each other as members of groups; economics, of the allocation of resources; political science, of governmental structures and functions. The slightest reflection shows that the definitions overlap.

History as the study of man's past embraces all his activities. There is, then, a history of science, a history of art, a history of economic doctrine, a history of sociology, and so forth. Are these to be regarded as a part of the study of each special discipline or as a part of the study of history? If the latter, historians would have to be specially trained in science, art, sociology, and so forth. If the former, there would be no subject matter left for historians per se, since every human activity is studied by a particular discipline. Another problem: a study of contemporary group activities and relations is classified as sociology, but what of a similar study about the Age of Augustus or of America just before World War I? Is it sociology or history? And what about the groups which are the basis of sociological study? There are political groups and institutions, economic groups and institutions, and so on. Are these not studied by political and economic science, respectively? What, then, is left for sociology? Just the residue after the other social sciences have chosen their fields? That would leave sociology only a few groups, like the family.

It becomes obvious that what distinguishes the social sciences from each other are the emphases and procedures of each discipline. Thus both political science and sociology are concerned with political groups and institutions, but constitutions, legislation, the formal structure of government, the problems of administration, are within the province of political science; and relations of status within political groups, the effect of government on community living, the social relations of political groups to other groups are within the province of sociology. Political science works on the materials of statutory law, charters, grants, constitutions, speeches, party platforms, and so forth; sociology works on the materials of political group behavior. Still, there is a real overlapping at some point, which we might call political sociology or political behavior. No intellectual scandal is involved in this. The natural sciences are in the same situation and at overlapping points usually use a joint name like biochemistry or astrophysics.

The subject matter of each social science is most readily discovered by reading chapter headings in standard works. Thus one finds in books on sociology such headings as social organization, social disorganization, social welfare, social change, social conflict, the family, class, and caste. And the emphases at the heart of each social science are found in its basic concepts. Economics constantly uses the concepts of exchange, price, supply and demand, distribution, competition, the market, equilibrium. Sociology functions with the concepts of group, society, socialization, class, caste, status, stratification, rank, mobility. Anthropology uses most sociological concepts but emphasizes, in addition, culture, acculturation, diffusion, pattern, race, endogamy, exogamy, totemism. Social psychology uses behavior, memory, perception, motivation, morale, suggestion, imitation. Political science employs the concepts of government and its forms—democracy, dictatorship, etc.—as well as parties, factions, power, representation, pressure groups, bureaucracy, administration, legislation, judiciary.[6] The list is incomplete but it shows the differences in the kinds of concepts employed in the several social sciences. These concepts are the ideas with which the social flux is arrested intellectually, the terms in which we classify and organize knowledge, the signposts which guide thought. A new concept offers a new way of seeing relationships and of organizing material; often it directs attention to what has been

[6] Many of these concepts are used in more than one science, although their origin and chief meaning are usually in only one. Concepts are borrowed both from other social sciences and from natural science. The concept of equilibrium in economics is borrowed from physics and chemistry, for example, and most of the social sciences have taken the concept of evolution from biology.

Borrowed concepts are sometimes used metaphorically, sometimes literally. When we say that bad government is the price of political lethargy, we are using "price" metaphorically, though clearly, and when we speak of the free market in ideas, we are using "market" as a metaphor. But when we speak of political motivation or social behavior, we are employing psychological concepts literally in the contexts of political science and sociology, for we are treating particular types of motivation and behavior.

neglected and so leads to discovery. The study of anthropology was redirected by the adoption of the concept of culture, and old explanations were discarded in favor of newer ones; so it is in all thought.

We have not listed the chief concepts of history because history has no special concepts. It is an inquiry unlike the others and must be treated separately. Although doubt has been cast on the scientific status of social sciences like sociology and political science, it is clear that they aspire to be sciences. But, although history uses rigorous methods to establish its conclusions, those conclusions are not general laws but particular "facts," and history seems to offer no possibility of scientific explanation. It does not imitate science in its structure and it does not have any concepts of its own.

What, then, is history if it is not a science? Perhaps it is a semi-scientific art, like biography. The method of history, according to some historians, is empathy—in this case, placing oneself in the position of men in the past as though they were actors on a stage, and then trying to understand their behavior from the "inside." We can agree that there is an element of art in the practice of historians, as there is in the practice of all scientists, and empathy of this sort is probably an especial virtue of historians. Also, much historical writing has the qualities of literary fiction. Characters are drawn so that they seem real, the settings are vivid and colorful, the large events of society give direction and meaning to individual life; in short, we are not asked just to understand the past, but to experience it, as we experience the incidents in a novel. This experience aids understanding by giving us concrete detail and by keeping all the elements of a time in proper proportion. The "first scientific historian," Thucydides, described diplomatic debates as though they were episodes in Greek drama, and thus gave his scenes immediacy and tension and his historical characters depth and reality.

Another possible description of history is that it has no subject matter of its own but is a *way* of studying any subject matter whatsoever. The past as such, it could be argued, cannot be the subject of any one particular science. It is too vast and, like the present, can be understood only by the use of all of man's separate studies. If we are interested in contemporary America, we study its government and laws, its economy, its social patterns, its arts, sciences, and philosophies. But if we are interested in ancient Rome, can we study only history? No, the argument runs, we must study the sociology, law, art, and so forth of ancient Rome. What, then, becomes of history? It deals with a different set of questions: not with a society at any particular time, but with the genesis and development of that society. Everything has a past and the study we call history is the approach to a subject through its past. History, thus conceived, is itself a principle of organization—the principle of chronology, of before and after, of growth and development—which may be used in any subject. So there is diplomacy and the history of diplomacy, science and the history of science, art and the history of art.

It would be enormously valuable if history were a science, for knowledge of general laws of history would benefit both social science and practical decisions. Is it impossible to attain such laws? Before we can answer this question, we must deal with another: Can history yield truth even about particular events? There are many who think not. Their argument is that history deals with the past: the past is over and done with—therefore history cannot test statements about the past by observation.

How can we answer this objection? Essentially, by pointing to the misunderstanding implied in it of the nature of historical "facts." The "facts before our eyes" when we study history are not events that occurred in the past and that have no existence in the present; they are rather such things as documents, diaries, coins, obelisks, statues, buildings. They exist in the present, and we use them as evidence that a hypothesis about past events is probably true or false. When we say "Columbus discovered America" we are stating not a "fact" in the sense of something that can be observed, but a hypothesis that, before proof, is an informed guess. The facts which are evidence of the truth of this hypothesis are existing records—lists of arrivals at and departures from ports, immigration documents (Columbus took Indians to Spain), public notices of the time, memoirs, and so on.

Where there is no evidence in the present, a hypothesis about the past remains speculation. But such speculation is frequently useful because it may lead to new discoveries in the present which can be used as evidence about the past. The death of the great English poet Christopher Marlowe, for example, was until recently cloaked in mystery. It was supposed that he had been stabbed to death in 1593 in a tavern brawl by a man named Francis Archer. But there was no satisfactory evidence that this was so; Archer's name was not mentioned in court proceedings, records of imprisonment, or lists of executed criminals of the time. A literary historian of our day, Leslie Hotson, speculated that there may have been no record of criminal proceedings because the killing of Marlowe was justifiable homicide, perhaps self-defense. The hypothesis fit the facts, but so would many others. Hotson reasoned that if he were right, evidence would be found where no one had yet looked, in royal pardons. He examined them and discovered a pardon made out to one Ingram Frizer for the killing of Christopher Marlowe in self-defense.

So we do not have to travel back in time to observe past events; we can infer them from present facts. And although "predict" seems a strange word to use in talking about the past, we do make predictions about the past on the basis of current knowledge and then, as in all science, we deduce from these predictions other propositions that are directly testable in the present or future. Hotson's procedure follows the general pattern of inquiry which characterizes all science.

Now we can deal with the question whether history can yield general laws. History, we have shown, can ascertain particular events in a scientific fashion. Why, then, should it not be able to generalize inductively from them? Here the objection is made that historical events are unique and that without observed recurrences we cannot generalize scientifically. But we have shown that the *same* event never recurs in nature or society; all we need for general laws are recurrent instances of the same *type* of event under instances of the same *type* of conditions. And we can find this in history as well as in any other social science. The Franco-Prussian War and the Crimean War are quite different in many ways, but they are both instances of wars. They can be used as evidence for a scientific law stating the conditions under which war takes place. We have no such law because it is difficult to isolate the relevant variables in dealing with an occurrence like war, which is not a single event but a complex of events.

Historians are usually at a loss when asked for a general law of history. Why is it so difficult to name one? Because when a law is formulated on the basis of historical evidence, it is usually classified as political, sociological, or economic. As we said, history is a way of studying subject matter, but that subject matter, itself divided into parts which lend themselves to somewhat independent study, makes up the material of the other social sciences. The social sciences have derived many hypotheses and laws from the historians. When economists study business cycles, for example, they turn to historical description of social events which document statements about business cycles in the past. When political scientists study war, when sociologists study migration, it is to history that they must go. And natural science, too, uses history: in observations of past eclipses, meteoric phenomena, and so on. One of the consequences of all this fishing in historical waters is that when the fish is landed he belongs to the fisherman, who is not himself a historian.

Even if history has no explicitly formulated general laws of its own, historians constantly assume the truth of laws of some sort—political, economic, sociological, psychological—in their attempts to explain. And no matter how much some of them protest that they are writing only a chronological record, historians *are* continually forced to explain. If, in the course of a chronological description, historians write about the fall of Rome, the American Revolution, or the murder of Robespierre, they select for their record those prior events which seem important, thus implying some causal connections. We have already discussed the necessity of general laws in explanation; here we need stress only the importance of making them explicit.

As an example (oversimple for effectiveness) of the ready way in which social laws are assumed in historical writing, consider the statement "The French Revolution came about because the people suffered intolerable conditions." First, the word "intolerable" both arouses emotion and begs the question. Conditions are defined as intolerable when people will not tolerate them, so of course if conditions are intolerable people will try to change them. Second, even if the specific conditions under which the French people revolted are stated explicitly, as they should be, but are then included under the word "intolerable," the word has perhaps added to a scientific description —"under conditions *a*, *b*, and *c*, the French people revolted"—an assumption that the behavior of the French was a *deliberate* response to these conditions, whereas the relation between the social conditions at the time of the Revolution and the psychological reasons for revolt needs explication. Also, historian and reader might be led to expect revolutions when people are most oppressed. And they would probably be misled, because people in the direst circumstances of poverty and slavery may busy themselves with the struggle to stay alive or solace themselves with religion, but they rarely rebel.[7]

If the assumption of deliberate action were made explicit it would be clear that it could not remain an assumption but would have to be proved. When it is not stated it is likely to be accepted implicitly, because the entire statement about intolerable conditions would have the persuasiveness of all tautologies not recognized as such.

To make it worse, if a general law is assumed, it is not at all clear which of several it is. Among other possibilities, the sentence we are examining might imply: (1) whenever the specified conditions of oppression exist (all other things being equal), there is a revolution; or (2) whenever there are such conditions *in France*, the people revolt; or (3) the French Revolution is unique in resulting from these conditions. The first is a general law which can be established only by an examination of all revolutions in history (or a fair sample) and the discovery of an invariant relation in them between the type of conditions described and revolution. The second is a general law which would require similar examination of revolutions in France and discovery of such a relation among the French. The third is not a general law; in fact, it denies that there is a law which is applicable to the French Revolution. This makes scientific explanation of the Revolution virtually impossible, as is always the case with an event regarded as unique. If we do not have a law stating the conditions of revolution, how can we possibly know which of the countless events prior to the French Revolution were the conditions that brought it about?

Not only does history rely on laws in the other social sciences but it helps provide those laws. For history is a storehouse of data which can be used as evidence. And the data may be extremely reliable because many methods of history are very advanced in the care with which all particular items of information are tested. In this sense, history is a scientific activity as well as a literary one, but it is perhaps not now a science because a science has general laws of its own. Yet basic to all science is its use of scientific method to warrant its conclusions, even if they are about particulars rather than generalizations. So we may think of history as a "descriptive science" somewhat of the

[7] Evidence for this is presented in James C. Davies, "Toward a Theory of Revolution," *American Sociological Review,* February 1962.

type of astronomy or anatomy. And there is no difficulty in distinguishing scientific history from that mixture of chronicle and legend which some groups or societies claim as their own history, or from that deliberate rewriting of the record to serve political advantage which the Soviet Union disseminates to its subjects and followers.

Still, history probably cannot be a science in the full sense of the word until there is a unified social science. And even then, its laws might be thought of as "belonging" to unified social science rather than to history itself. Yet, however we classified those laws, they would enormously increase our knowledge and could yield unequaled social control.

No matter what its status as science, history is indispensable to the other social sciences and they are indispensable to history. New theories in sociology and political science have led historians to look for relationships they had neglected before, and psychoanalysis has brought new interpretations of important historical figures. Further, social science is now doing work which will enable the future historian to deal more exactly with our own day. After all, histories are full of statements like "There was great dissatisfaction with this measure." How does the historian know? Because people acted in a way that he interprets as the result of dissatisfaction. But he may very well be wrong, or be loose in his use of the word "dissatisfaction." [8] In addition, he may not be aware of some dissatisfactions at all, because they are not always expressed in action. The social surveys and opinion polls of our own time, however, will allow the historian to know more accurately whether people were dissatisfied, and in what proportion to the population, what group or background they came from, and the details of their dissatisfaction.

More striking, perhaps, than the reliance of history on the other social sciences, is its use of advances in natural science. This may bring about the actual confirmation or disconfirmation of particular historical hypotheses. The most dramatic instances are found in recent changes in historical accounts of what used to be regarded as legendary and mythical material. There is the story of the Flood, for example,

[8] Whatever diaries, letters, or other personal expressions are at his disposal, it is difficult for the historian to know how representative they are, how adequate in their observations of others, and how much they were a response to private, rather than social, events.

as related in the Bible and in other ancient documents. The Flood is interpreted Biblically as a visitation of God's wrath on man for his sins. The very description of the event might be sheer invention, skeptical historians thought, or a sweeping metaphor of God's anger at man's immorality. But historical conclusions in such matters depend on geology and paleontology, which discover and date important changes in the surface of the earth, and these have by now provided ample evidence that there was at least one great flood at about the time specified by the ancient accounts.

Interesting changes in historical acceptance result also from work in archaeology. Nineteenth-century historians were so skeptical of the legends of the ancient Greeks that they would not accept them as descriptions of actual events. So when Grote wrote his famous history of Greece he totally discounted the Trojan War, which was Homer's subject in the *Iliad,* and assumed that there was no historical reality corresponding in any way to such stories as that of Theseus and the Minotaur. But there were archaeologists less skeptical, more impressed with the historical basis of the *Iliad* and of Greek legend. Heinrich Schliemann went to what was then Turkey and dug up Troy; he found, in fact, nine Troys, one on top of another, of which Troy VI corresponds in all essentials to Homer's description.

In 1900, J. B. Bury published what seemed to be a virtually definitive *History of Greece to the Death of Alexander the Great.* In the same year, Sir Arthur Evans started his diggings in Crete, finding so many important things that he was still at it twenty-five years later. In the ancient capital of Knossos, Evans found the great palace of Minos, whose plan is that of a labyrinth, very much like the legendary one in which Theseus, aided by Ariadne's thread, found and slew the Minotaur. Bury's book, therefore, was obsolete in its first part almost as soon as it appeared. In his new edition, Bury wrote as preface: [9]

> The excavations of Sir Arthur Evans at Cnossus began in the year in which the first edition of this History of Greece appeared. . . . His amazing discoveries there, followed and supplemented by the work of other explorers on many other prehistoric sites in Crete, have transformed our knowledge of the Aegean civilisation of the second millennium, and placed in a new focus the problems of early Greece. In consequence of these discoveries, and of other researches (among which I may mention

[9] Second Edition, 1913, published by Macmillan and Co., Ltd., London.

especially Professor Ridgeway's *Early Age of Greece* and Mr. Leaf's *Troy*), it has been necessary to rewrite the greater part of Chapter I. An account of Cretan civilisation is included; the view that the pre-Achaean inhabitants of Greece were not Greeks, which it seems to me no longer possible to maintain, is abandoned; and the Trojan War is recognised to be an historical event.[10]

History, much more than natural science and even more than the other social sciences, is subject to a lack of objectivity stemming from the investigator's own values. The most obvious partisanship is found in the outright distortion of history due to national, ethnic, and religious prejudice. History has been used by the modern nation-state as a way of creating in its children attitudes favorable to national solidarity and survival. There has been widespread recognition of this. After the Franco-Prussian War of 1870, for example, a common remark throughout Europe was that the victory was won by the Prussian schoolmaster. For years history texts, especially in the primary and secondary schools, bore the academic burden of instilling patriotic sentiment in students.

In a study by Arthur Walworth called *School Histories at War*, history texts used in the United States were compared with those of England, Canada, and Mexico in their accounts of American wars against those nations. The differences, even in matters of simple "fact," were glaring and ludicrous. Both Americans and Mexicans claimed victory in the Battle of Buena Vista, and American and English treatments of the Battle of Bunker Hill make it seem like two separate events. Common to most of these books is the use of a national enemy as a way of creating patriotic fervor, much as there are "traditional rivals" of schools and athletic clubs—Yale and Harvard, Army and Navy, Michigan and Minnesota. American schoolchildren have had to outgrow hatred of the Redcoats in order to cope with the realities of current international politics. Canadians, who have had less strife with other nations, have used Americans as a national enemy, writing in history texts of their arrant cowardice, and preserving as national monuments eighteenth-century forts with antique cannon

[10] Further corroboration of Homer has more recently been provided by the decoding of one type of ancient Cretan writing, the Minoan linear B script. This turns out to be Greek written in a Minoan script, and some of it contains the names of Homeric heroes.

pointed at the American border. The American Legion has revealed persistence of this spirit in a charming euphemism: they once issued a statement urging that American history be taught "optimistically."

Other evaluations by historians enter more subtly, sometimes in a single word, and an incautious reader can easily be led to accept a value that is not made explicit. Consider again one of the first pieces of historical information imparted to every American schoolboy: Columbus discovered America. Can a place be discovered when people already live there? Why do we not say that the inhabitants discovered it? Perhaps they always lived there, and a place can be said to be discovered only by people who originally came from elsewhere. But although American Indians probably came from Asia, no one says that the Indians discovered America. Do we mean that only white men, or Europeans, can discover a place? If so, let us remember there is some evidence Vikings came to America before Columbus did. Obviously, the word "discovered" has an evaluative charge. We approve of people who discover things and we approve of ourselves when discoveries are made by members of our own group. What is probably implied in the meaning of "discover" in this context is that discovery of a place is attributed to the leader of an expedition which brought knowledge of that place back to its own society.

Discovery need not be geographical in order to be honorific. National pride in technological, scientific, and artistic discovery or innovation is evidenced by the conflicting claims of historical accounts in different countries. The invention of the steamboat is attributed in English texts to Symington and in French texts to Jouffroy, but Americans "know" that the steamboat was really Robert Fulton's invention. Americans can be impartial about the discovery of the calculus, independently arrived at by the Englishman Newton and the German Leibnitz, but they are usually clear that the Wright brothers and not Langley or anyone else invented the airplane. Russian nationalism has led to Soviet claims that Russians discovered penicillin and invented the airplane, the radio, and many other things.

Evaluations are also implicit in historical classifications. If we ask, "Who were the people of Europe in the twelfth century?" we may be answered with "Christians, Jews, and Mohammedans" or "Frenchmen, Spaniards, Italians, and so forth," or "Latins, Teutons, Slavs, and so forth," or "Nordics, Alpines, and Mediterraneans." Each is a correct answer in that it provides a complete classification of Europeans at the time. The one we accept reveals our interest in religion or nationality or ethnic background or anthropological type as a principle of classification. For some purposes of description or analysis one is actually better than another. But sometimes a classification is chosen because the category it expresses simply seems "more important," although not for any particular purpose. We may just feel that religion is more important than nationality, or vice versa; what we mean is that for *other* nonscientific purposes one of the two is more valuable.

A more serious instance, because the unwary are prone to think that a particular classification is correct rather than merely useful, occurs when a classification is changed. Histories used to classify the stages of "prehistoric man" as nomadic, hunting and fishing, and agricultural. Now they classify the stages of "preliterate man" as Old Stone Age, New Stone Age, and so on. The change is probably evidence of a shift in cultural emphasis. It used to be felt that nothing about man was more important than his occupation; now it is felt that the significant thing is the kind of tools he employs. Yet if we do not understand that the new classification results from a shift in perspective or evaluation, we may believe that the old classification was false and this one alone is true.

Every new thing we learn about man and society, every new social evaluation of what is important, each succeeding philosophy of society and history affects the way we write history. When we adopt a new theory we select different matters out of the welter of past events and discard matters that were important for older theories. Human behavior, for example, is interpreted differently from the standpoint of a philosophy of individualism than it is from the standpoint of a collectivist philosophy. For the former, man is chiefly a person, society is nothing more than the network of man's relations to his fellows, and each society is appraised in terms of what it does for and to individual men. For the latter, man is chiefly a social unit, society is an "organism" of which individual men are only "members," and each man is appraised in terms of what he does for and to his society. The differences in these philosophies make for differences in histories based on them. Individualist histories may contain much about the effects of social change on individual lives, and collectivist histories may emphasize social organization and law. If a historian believes, with Thomas Carlyle, that ". . . Universal History . . . is at bottom the History of the Great Men who have worked here," he will write very differently from the way he would if he believed, with Marx, that "the law of history is the law of the class struggle."

A new social philosophy is essentially a combination of new perspectives (ways of looking at things, emphases, frames of reference), new methods, and new or differently ordered values. It is not itself testable, although it will probably include many empirical statements which are testable. But these are incidental; the importance of a social philosophy is in the help given to social thought by the new world view, the way in which we understand ourselves, others, and society, and are led to new discoveries. Two philosophers of our own time, the American George Herbert Mead and the German Ernst Cassirer, wrote of man as a symbol-making and symbol-using animal. Their writings led to new emphases in social psychology and sociology on the importance of language and linguistic usages, on the elements of symbol and language in the process of socialization, on the transmission of the culture through symbols, on the web of communication and symbol which holds society together. The new emphases brought new hypotheses supported by evidence hitherto neglected or unnoticed, and entire texts in social psychology are now written from the standpoint of communication and the uses of symbol.

All written and oral history is based on some kind of philosophy, not necessarily a technical one associated with a great thinker, like Aristotle or Kant, but a somewhat consistent view of the world and of society. Not only social philosophies but religious philosophies, like Christianity, and scientific philosophies, like mechanism, have shaped visions of history. Even when the historian thinks he is merely recording events chronologically, without attempting explanation, he is necessarily selective about the events he records, and selectivity implies a principle of selection. That principle usually contains some ideas about explanation: what sorts of events bring change, what are the most important kinds of social activity (that is, in terms of which the others can be explained), and so on. The principle of selection alone involves a philosophy. As old philosophies are discarded and new ones take their places, history is rewritten; so it always expresses the meaning of the past to the present or to some aspect of the present.

# 6

# The Role of Science in Society

It took many years of civilization for science as we know it to come into being. Of course, men have survived, and even lived well, on the basis of a universal rudimentary science which is a combination of prudence and untrained observation. The Greeks and medievals practiced science sporadically, and sometimes brilliantly, but it was not until Galileo and Kepler that mathematical reasoning, precise observation, and experiment were joined in a systematic and continuing enterprise.

Why did it take so long for science to develop? Why has it never developed among primitive peoples? Not because they were intellectually incapable of discovering it, but because there were rival ways of attaining truth that seemed to offer the knowledge and assurance men wanted—religion, magic, secular tradition, and, later, pure reason—and these fought off other claims to truth. Men were not raised in an intellectual vacuum but were conditioned to belief, and even if science had existed as a competing claim, the psychological attractions of magic and religion might have decided the issue in their favor.[1] It takes a special sophistication to return to innocence, or simplicity, and

[1] The Moslem general who burned the great library at Alexandria which contained the works of Greek civilization argued that all truth was in the Koran, that if the books in the library contained that truth they were unnecessary, and that if they contained anything else they were false. Book

to accept what is before every man's eyes as the foundation of knowledge.

The basic social condition for the existence of science is an attitude, reasonably widespread through a society, which accepts observation as the only way of testing factual truth. To be sure, men have always learned from observation and have even experimented under very primitive conditions. An experiment, after all, only provides an occasion for observation which we might not otherwise be able to make, or provides a better occasion than otherwise exists. Often the observation requires comparison of what happens when we intervene in natural processes with what happens when we do not. The primitive man who generated heat when rubbing two pieces of wood together may not have intended an experiment, but his action became an experiment when he compared (implicitly) the result of friction with the result of leaving the two pieces of wood alone. Scientific experiment differs in two ways from primitive experiment. It is conceived as a result of, and executed as a test of, hypotheses; and it has a much higher degree of deliberate control. Primitive observation and experiment led to practical knowledge of particulars and to craft, not to general laws and science.

When men first discover the power of reason, they are likely to use it as a method in itself, or as an elaboration of what they already believe without it. An ancient Greek philosopher, Parmenides of Elea, argued with the greatest cogency that there was no motion. To simple people who answered that they saw things move Parmenides replied that it was all illusion and that he could prove rationally that nothing *really* moved. And other superb minds, among them the scholastic philosophers of the Middle Ages, tried to solve problems that we now think of as scientific by deduction from theological premises or by citing authorities. Galileo was bitter about such a view of science. He wrote, "As I wished to show the satellites of Jupiter to the Professors in Florence, they would see neither them nor the telescope. These people believe there is no truth to seek in nature, but only in the comparison of texts."

burning, whether it is done through the dictates of a political party, a religion, or a nation, is always a symptom of fear and hatred of questioning, speculation, or science.

Even when science has a history of great intellectual victories, as it does today, almost all the older ways of arriving at truth persist and science is at times still threatened by despots and demagogues. But, one might ask, if there are still rival claims to truth, why do their protagonists not argue the case and test their claims against the claims of science, instead of merely using power? The answer comes when we ask another question: What sort of test could be used? If it is the sort that science uses, then of course science will win. If it is the sort the rival uses—the will of a Leader, the answer of any other oracle, the statements in *Das Kapital, Mein Kampf,* or any other scripture—then of course the rival will win. If the kind of test could be agreed on, there would be no disagreement in the first place, for the disagreement is precisely on how ideas should be tested.

Rival claims can coexist peacefully with science when the areas of each are seen to be different. The claims of science as a way to truth pertain only to matters that can in the end be tested by observation. To be sure, some scientific philosophers have denied that there is any truth outside science, but that is a debatable contention, not often made by scientists. And to a great extent the plausibility of the contention depends on restricting by definition the meaning of the word "true" to propositions that have been tested by observation, thus begging the question whether only such propositions are true. Few people, even among those insensitive to activities that are not scientific, would deny that nonempirical matters are important or valuable. And there are scientific philosophers who accept other claims to truth, claims in nonscientific matters; they show their acceptance, for example, by their religious belief.

Yet the definition of the areas appropriate to different subjects must be in terms of logical content, not in traditional or institutional terms, or the claims to truth will again conflict. Medicine does not include the existence of God in its field of competence, and religion should not compete with medicine by dealing with the causes of disease—attributing disease, perhaps, to sin or lack of faith. In particular, it has traditionally been held that a study of man and his works, especially his values, belongs to the arts and religion, not to science. This claim must be qualified or there will be conflict with social science. Just as religion gave up specific areas to natural science—astronomy and agriculture, for example—so art and religion must give up some areas to social science. But this does not weaken art and religion in the long run; it strengthens them by purifying their subjects of extraneous elements and allowing concentration on what is truly relevant to them.

The systematic study of society is a much later development in history than the systematic study of nature. The revolutions of the spheres were charted long before business cycles were, and the speed of falling bodies was calculated earlier than the rate of suicide. There was the beginning of natural science in ancient Greece. An accurate prediction of an eclipse of the sun (according to some students, May 28, 585 B.C.) has been attributed to a Greek philosopher, Thales of Miletus. But Greek social science—what there was of it—was incidental to political and ethical theory. Modern natural science came into being fully in the sixteenth century and had great triumphs in the seventeenth. Modern social science came into being fully in the eighteenth century and developed slowly through the nineteenth. And twentieth-century social science is, of course, far behind natural science in the breadth of its discoveries and the rigor of its methods.

On the face of it, the disparity in time between the development of natural science and that of social science is startling and seemingly paradoxical. After all, some knowledge of society is indispensable to every man, and society is there, right around him, to be studied; but we can live without knowledge of the stars, which are very far away. How is it, then, that a science of astronomy was developed earlier than a science of society? Probably some men assumed they knew what was directly under their noses, so that "common sense" and practical wisdom were sufficient for understanding society, while the distant stars obviously had to be studied. Also, many men believed that the stars guide our destinies, so knowledge of the stars was expected to yield more practical information than could be garnered from knowledge of society. Then, too, the belief that man has absolute free will led some people to conclude that human behavior cannot be explained. Even more important, the questions men asked about society were answerable in terms of practical politics, simple prudence, traditional wisdom, and theological morals; they were not answerable in terms that would lead to science. And men constantly confused questions about nature with questions about society and morals.

Again and again, the question was asked, "Why are we suffering famine?" and the answer came, "As a punishment for our sins." In

Sophocles' *King Oedipus,* the city of Thebes suffers from a plague. An oracle is consulted and says that a sinful man lives in Thebes and that the plague will continue as long as he remains in it. The oracle's answer is assumed throughout the play to be correct. The sufferings of the Old Testament Jews were attributed directly to the wrath of the Lord because of their sins. Jeremiah, for example, explained a drought: ". . . thou hast polluted the land with thy whoredoms and with thy wickedness. Therefore the showers have been withholden, and there hath been no latter rain."

The meaning of a question is determined by the kind of answer that is demanded, and these ancient questions about famine, drought, and plague were questions of theological morals, not of scientific agriculture or medicine. Today we might answer the question, "Why are we suffering famine?" with a statement like "Because we haven't practiced crop rotation." But such a proposition would not have been an answer to people whose question meant, "What evil have we done that offended the gods?"

Science does not necessarily come into existence when an earlier claim to truth loses its hold on a community. The vacuum of belief may be filled by still another claim, even of the same type. One religion may replace another religion, or one authority replace another authority. Probably the existence of socially approved motives and goals which can be served especially well by science is a condition for its birth.

Modern science began in the Renaissance, and one of the highly approved goals of the Renaissance was power. "Knowledge is power," said Francis Bacon, prophet and propagandist of empirical science, meaning that science is the means to the goal. Most of Bacon's contemporaries felt as he did, although some scientists—Galileo and William Harvey, for instance—may have been more concerned with knowledge itself, which is a direct aim of science, than with power, which is one of the uses of science. The late Middle Ages and the Renaissance often equated the scientist with the magician because both were thought to be essentially concerned with power, and most people did not see the difference between power as a result of natural means and power gained through traffic with the supernatural. Roger Bacon was thought of as a magician, and the great story of Faustus, who sold his soul to the devil for the sake of knowledge (partly because knowledge gave him power), seemed a proof that magic and science were one.

Modern social science, too, began in the Renaissance, although one or another aspect of social science can be found as far back as ancient Greece. Certain attitudes and beliefs seem to have been required before social science could be practiced, and these did not exist sufficiently in classical civilization. Whatever the relation between society and ethics, society has to be studied as it is, not as it ought to be, if we are to have factual knowledge of it. In the end, a social scientist may as a man approve or condemn any society he has studied, but he must first, as a scientist, know what that society is like. To attain that knowledge he must study it impartially. Older views of a particular social order as divine, or as the expression of the will of the gods, made full-fledged social science almost impossible.

Coupled with theological and ethical concerns were legal ones. Social relations were treated in terms of legal norms, and the state was thought of as embracing all of society. Social thought in the ancient world was at one and the same time ethics, political and legal theory, sociology, and economics. Plato's *Republic* is a superb example of this combination. But some purposes of science were served only when each of these studies was separated from the others and pursued independently. Now that each social science has been autonomous for some time, it is important to relate their conclusions so as to create a unified social science. But it was impossible to begin with unification.

Political science was the first social science to be developed, perhaps because it most obviously dealt with power, that vital theme of the Renaissance. Traditionally, the origin of the science of politics is ascribed to Niccolò Machiavelli, whose small book *The Prince,* published in 1532, became a handbook for rulers on the attainment and maintenance of political power. Machiavelli was deeply concerned with the establishment of a single political power to unify Italy and was prepared, if necessary, to accept unification by conquest. *The Prince* contains his observations on the way people in fact behave in politics and his advice on ways to wield power most effectively. There are a number of moral comments in *The Prince,* but on the whole Machiavelli separated the study of politics from its older context of the moral and judicial, thus creating the basis for a scientific study of political behavior. Machiavelli's statements, however, were quickly converted to a philosophical use which seemed justified by their cynicism about man.[2] They became the foundation of a systematic theory of absolute monarchy worked out by the seventeenth-century English philosopher Thomas Hobbes, who may have been the first man to write of his hope for a complete science of human nature on the model of physics. Human science, he probably assumed, would justify his own political theory more fully than the personal observations of Machiavelli did.

But it was not until the eighteenth century that psychology and economics started to fulfill Hobbes's vision. Both studies developed in England and France in an atmosphere of philosophical empiricism—the belief that experience, in the form of physical sensation, is the origin of all our knowledge and belief. It seemed to follow from empiricism that the differences in men were due to differences in their experience or environment, and therefore that men were equal at birth. Why, then, eighteenth-century thinkers asked, do men have unequal opportunities and stations in life? Can this be just? If not, how can it be remedied? And their answer was: by a redistribution of social power. So the attempt to create a science of man by discovering the laws of his nature and the laws of his social behavior, with no reference to what man or society ought to be like, remained bound to value judgments which were, of course, not scientific at all. And from a defense of monarchy, social science became a defense of democracy, even a call to revolution in its behalf.[3]

[2] Machiavelli wrote: ". . . It may be said of men in general that they are ungrateful, voluble, dissemblers, anxious to avoid danger, and covetous of gain; as long as you benefit them, they are entirely yours; they offer you their blood, their goods, their life, and their children . . . when the necessity is remote; but when it approaches, they revolt. . . . And men have less scruple in offending one who makes himself loved than one who makes himself feared; for love is held by a chain of obligation which, men being selfish, is broken whenever it serves their purpose; but fear is maintained by a dread of punishment which never fails."

[3] The eighteenth century has been called the Age of Reason because many of its thinkers tried to brush aside tradition and come to conclusions based on what the human mind could discover, starting from scratch. *Rationalism*—the belief that everything could be known by deduction from self-evident axioms—was a seventeenth-century aspiration which still lingered, but the great social and political themes, culminating in the Declaration of Independence and the Rights of Man, were based on a concept of reason that substituted sensation for axioms as the base of thought.

The basic assumptions of eighteenth-century social science were bound up with individualism and rationality. Sensations, from which came all the furnishings of the mind, were obviously individual and unsharable; I can tell you about my pleasures and pains, but they are mine alone. Rationality was understood as a calculation of advantage, because men were supposed to be dominated by a natural desire to improve their lot, by a drive to pleasure and an aversion to pain. Being rational consisted in attaining these inescapable ends with the greatest efficiency and economy. Eighteenth-century assumptions thus lent themselves to individualism in psychology, *laissez faire* in economics, and representative government in politics. And the very leaders of the social sciences had purposes beyond the desire to understand man, purposes of a moral, philosophical, or political kind. Adam Smith, the founder of classical economics, was a professor of moral philosophy, and his great work in economics, *The Wealth of Nations,* was to him continuous with his earlier book, *The Theory of Moral Sentiments.* Even historiography had social purpose. David Hume, philosopher, economist, historian, wrote of history, "Its chief use is only to discover the constant and universal principles of human nature."

Although the social sciences have by our time overcome their early assumptions and are pursued more directly for the sake of knowledge, keeping fairly separate the question of their use, one heritage of the eighteenth century remains—contemporary social scientists are often reformers at heart. Perhaps reformers are attracted to social science, so that they can learn how to bring about the changes they desire; perhaps the study itself makes men into reformers. There is an easy transition from learning, for instance, that the factual arguments used in defense of lower status for Negroes are false, to advocacy of equality for Negroes.[4]

This reformist zeal refutes a common charge that social scientists are deterred from conclusions they would otherwise come to by fear of the way they would be treated. After all, the argument runs, the social scientist is in a very different social situation from the natural scientist. New discoveries in chemistry or physics are acclaimed and the discoverers are knighted, awarded the Nobel Prize, lionized. There are

[4] A teacher of sociology and anthropology at an Ivy League college has referred to Joseph Conrad's novel as *The Negro of the Narcissus,* unconsciously repressing the real title because it contained a word that offended him.

no vested interests which are affected by such matters as whether the universe is expanding or contracting, so the conclusions of natural science can be treated with impartial respect. But there are interests which are affected by social conclusions—for example, whether a capitalist economy can in our time continue successfully or whether a transition to socialism will attain economic goals better. Society is not likely to accept the notion that a falling birth rate can be overcome by giving up monogamy for polygamy, and it may persecute the man who proposes such a change or merely publicizes the point.

There is a seeming plausibility to the argument that social scientists respond to the pressure of interest groups. And on the basis of radical premises about the behavior of "the capitalist class," social science appears to be an apology for the existing order.[5] But there are overwhelming arguments—in addition to the fact we have noted, that social scientists do often hold unpopular views—to show that social scientists are remarkably free in democratic countries. First, many social scientists who have come to liberal, even radical, conclusions have—almost as much as natural scientists—been highly honored. Among the more distinguished names that come to mind some of the best known are those of Lord Keynes, Lord Beveridge (neither of whom inherited his title), John Dewey, and Gunnar Myrdal. Second, the reputations of social scientists, like those of natural scientists, depend basically on the judgment of their colleagues. Most work in social science is too technical to be understood, and often too boring to be read, by the layman. And third, in so far as a man sticks to his last and writes social science, not propaganda or political tracts, his colleagues in democratic countries show little or no tendency to judge him by the possible consequences of his ideas for social and political action.

So it is rarely the threat of overt force or loss of reputation that dictates the thinking of social scientists. What is hard for them to overcome is the unconscious adoption of the values of their own society, which they resist little more than other men. In the past, before the prestige of science was so great, there were many more social pressures brought to bear on the scientist. But it was the natural scientist who ran the greatest risks, because many popular beliefs about nature were

[5] Read any official Communist account of science, especially social science, in capitalist countries, or an unofficial account, like Barrows Dunham's *Giant in Chains,* Little, Brown, 1953.

derived from theological (and even political) premises, and these were sacrosanct in a way that economic views, for example, never were. Thus, in the seventeenth century, natural scientists could be crushed by the pressures of an entrenched orthodoxy. We can gain perspective on the relationship between scientific novelty and rival claims to truth about factual matters by recalling the quarrel of the Inquisition with Galileo.

Galileo was forced by the Inquisition in 1616 to retract the belief that (in the official words of the Inquisition) "the sun is immovable in the center of the world, and that the earth moves, and also with a diurnal motion." Yet in 1623 he published a book, *Dialogues of Galileo Galilei on the Two Chief Systems of the World*, in which he compared the Ptolemaic and the Copernican systems of astronomy to the advantage of the Copernican. To make matters worse, Galileo had a character named Simplicius utter some remarks that had originally been made by the Pope. The Inquisition ordered the old Galileo (he was seventy and going blind) from Florence to Rome for trial. The result was a document of condemnation by the Inquisition and a public recantation by Galileo.

Admitting that Galileo held that the Copernican system was "merely probable" and adding "which is equally a very grave error" (doubtless because laws of nature should be certain), the Inquisition pronounced sentence:

> Invoking, therefore, the most holy name of our Lord Jesus Christ, and of His Most Glorious Virgin Mother, Mary, We pronounce this Our final sentence. . . . We pronounce, judge, and declare, that you, the said Galileo, by reason of these things which have been detailed in the course of this writing, and which, as above, you have confessed, have rendered yourself vehemently suspected by this Holy Office of heresy, that is of having believed and held the doctrine (which is false and contrary to the Holy and Divine Scriptures), that the sun is the centre of the world, and that it does not move from east to west, and that the earth does move, and is not the centre of the world; also, that an opinion can be held and supported and probable, after it has been declared and finally decreed contrary to the Holy Scripture, and consequently, that you have incurred all the censures and penalties enjoined and promulgated in the sacred canons and other general and particular constitutions against delinquents of this description. From which it is Our pleasure that you be absolved, provided that with a sincere heart and unfeigned faith, in Our presence, you abjure, curse, and detest, the said errors and heresies, and every other error and

heresy, contrary to the Catholic and Apostolic Church of Rome, in the form now shown to you.

But that your grievous and pernicious error and transgression may not go altogether unpunished, and that you may be made more cautious in future, and may be a warning to others to abstain from delinquencies of this sort, We decree that the book *Dialogues of Galileo Galilei* be prohibited by a public edict, and We condemn you to the formal prison of this Holy Office for a period determinable at Our pleasure; and by way of salutary penance, We order you during the next three years to recite, once a week, the seven penitential psalms, reserving to Ourselves the power of moderating, commuting, or taking off, the whole or part of the said punishment or penance.

The abjuration demanded by the sentence was signed by Galileo:

I, Galileo Galilei, . . . swear that I have always believed, and with the help of God, will in future believe, every article which the Holy Catholic and Apostolic Church of Rome holds, teaches, and preaches. But because I have been enjoined, by this Holy Office, altogether to abandon the false opinion which maintains that the sun is the centre and immovable, and forbidden to hold, defend, or teach, the said false doctrine in any manner; and because, after it had been signified to me that the said doctrine is repugnant to the Holy Scripture, I have written and printed a book, in which I treat of the same condemned doctrine, and adduce reasons with great force in support of the same, without giving any solution, and therefore have been judged grievously suspected of heresy; that is to say, that I held and believed that the sun is the centre of the world and immovable, and that the earth is not the centre and movable, I am willing to remove from the minds of your Eminences, and of every Catholic Christian, this vehement suspicion rightly entertained towards me, therefore, with a sincere heart and unfeigned faith, I abjure, curse, and detest the said errors and heresies, and generally every other error and sect contrary to the said Holy Church; and I swear that I will never more in future say, or assert anything, verbally or in writing, which may give rise to a similar suspicion of me; but that if I shall know any heretic, or anyone suspected of heresy, I will denounce him to this Holy Office, or to the Inquisitor and Ordinary of the place in which I may be. I swear, moreover, and promise that I will fulfill and observe fully all the penances which have been or shall be laid on me by this Holy Office. But if it shall happen that I violate any of my said promises, oaths, and protestations (which God avert!), I subject myself to all the pains and punishments which have been decreed and promulgated by the sacred canons and other general and particular constitutions against delinquents of this description. So, may

God help me, and his Holy Gospels, which I touch with my own hands, I, the above named Galileo Galilei, have abjured, sworn, promised, and bound myself as above; and, in witness thereof, with my own hand have subscribed this present writing of my abjuration, which I have recited word for word.

At Rome, in the Convent of Minerva, June 22, 1633, I, Galileo Galilei, have abjured as above with my own hand.

The story is told that when Galileo finished reciting his abjuration, he muttered: "*Eppur si muove*" ("And yet it does move"). But Bertrand Russell is probably right in his comment: "It was the world that said this—not Galileo."

The most serious controversy in twentieth-century America between science and society involved natural, not social, science.[6] Religious interests concerned with the nature and genesis of man and the universe have been much more powerful opponents of science than, for example, economic interests concerned with the nature of society. And some social opposition has been created by the folly of scientists who substitute ill-thought-out reform for science itself or make preposterous metaphysical, psychological, and theological claims, for example, that man is really a machine, that love is only a matter of chemistry, that the universe is a vast accident, that the one thing worthy of human worship is Reason, or Science. Part of the nineteenth-century attack on evolution was inspired by such wild claims, which were ridiculed by a number of people, including the great British prime minister Benjamin Disraeli. It is worth quoting from one of Disraeli's novels, *Tancred,* published in 1847, twelve years before evolution became fully scientific with the publication of Charles Darwin's *Origin of Species:*

After making herself very agreeable Lady Constance took up a book which was at hand, and said, "Do you know this?" And Tancred, opening the volume, which he had never seen, and then turning to its title-page, found it was *The Revelations of Chaos,* a startling work just published, and of which a rumour had reached him.

"No," he replied, "I have never seen it."

"I will lend it you, if you like; it is one of those books one must read. It explains everything, and is written in a very agreeable style."

[6] The Scopes Evolution Trial, Dayton, Tennessee, July 1925.

"It explains everything!" said Tancred; "it must indeed be a very remarkable book!"

"I think it will just suit you," said Lady Constance.

"To judge from its title, the subject is rather obscure," said Tancred.

"No longer so," said Lady Constance. "It is treated scientifically; everything is explained by geology and astronomy, and in that way. It shows you exactly how a star is formed; nothing can be so pretty! A cluster of vapour, the cream of the milky way, a sort of celestial cheese, churned into light. You must read it; 'tis charming."

"Nobody ever saw a star formed," said Tancred.

"Perhaps not. But you must read the *Revelations;* it is all explained. But what is most interesting is the way in which man has developed. You know, all is development. The principle is perpetually going on. First there was nothing, then there was something; then I forget the next, I think there were shells, then fishes; then we came. Let me see, did we come next? Never mind that; we came at last. And the next stage will be something very superior to us; something with wings. Ah! that's it: we were fishes, and I believe we shall be crows. But you must read it."

"I do not believe I was ever a fish," said Tancred.

"Oh! but it's all proved! By geology, you know. You see exactly how everything is made: how many worlds there have been; how long they lasted; what went before, and what comes next. We are a link in a chain, as inferior animals were that preceded us; we in turn shall be inferior; all that remains of us will be some relics in a new red sandstone. This is development. We had fins: we may have wings."

To return to our theme—the desire for power over nature and man was an important factor in the origin of science and its initial acceptance by society. Science has continued to receive social support partly because of its power, and it has indeed had many uses: natural science for technology, and social science for business, government, and education.

Science yields control over nature only indirectly, through technology. Technology has not only changed the face of society—buildings, transportation, communication—but has affected its mores and institutions. Thus natural science indirectly worked a great change in society, and the astrologers (though for the wrong reasons) and Francis Bacon were right in thinking that knowledge of the stars and of matter would yield social power. The physicists have even unleashed the force of the atom and the engineers are harnessing it. But social power has been used by men largely for individual purposes, and the great social changes wrought by technology have been accidental, not deliberate. Men have been changing society unthinkingly, without the opportunity to decide whether or not they wanted the changes.

Suppose a supernatural power had offered us a magic carpet which would get us from place to place with enormous speed, but demanded in return the annual sacrifice of thousands of men and women. Would we have accepted the gift? Probably not. Yet we have accepted its equivalent, the automobile, accepted it without being asked and without reflection. During the period of our participation in World War I there were more deaths from automobile accidents in the United States than were suffered by the entire A.E.F. in battle, and at present the annual figure for death by automobile is about 38,000, while injuries exceed 3,000,000.

In contrast, the use of social science to change society could be deliberate. What we know about the causes of business cycles, for example, can be used directly to avoid or minimize economic depression. The harm done by unplanned technology—not only to the bodies but to the minds of men—is not overcome by still more unplanned technology; if it can be overcome, or minimized, it will be by use of the knowledge provided by the social sciences. We can use knowledge of the conditions of daily social life and human responses to changes in it, for example, to alter our traffic routes, create psychological tests for drivers, even replan our cities.

Yet it is folly to be overoptimistic, like H. G. Wells, about our scientific future, for power is always dangerous. As we know more and more about how men act and how to control their actions, we are more able to enslave as well as to help. Electric shock therapy, useful in treating psychological depression, can impair or destroy memory. The prefrontal lobotomy, an operation supposed to make the incurably insane cheerful and tractable, can make men unthinking and obedient. The techniques developed to make advertising effective could be the chief mold of belief if used by a totalitarian society which offers no access to knowledge. Technology itself imposes no necessity that it be used for good rather than for ill—the blade of a knife is indifferent whether it cuts bread or throats. Modern technology yields a power which, misused, may enslave or destroy as never before. So concern with the moral and social context in which technology is used becomes more important with every technical advance, and politics, in which social decisions are carried out, allows less and less room for mistakes.

But these considerations lead us from conditions for the existence of science to conditions favorable and unfavorable for science once it does exist. We shall see that in a society where science is useful and has already proved its usefulness it is not eliminated even though the proponents of rival claims to truth wield social and political power. But it may be hampered and distorted.

Some conditions necessary for the existence of science are also favorable to its development when they are present in marked degree. Rationality and individualism especially, when they are widespread and highly valued in a society, tend to give science great social approval. And the approval of innovation which characterizes business activities helps make scientific discovery honored and rewarded.[7] Less obvious conditions for scientific advance are in the responses of society to the values internal to science.

Science as product, as a set of statements about the world, neither contains nor implies judgments of value. But science as an activity does. The conduct of members of the scientific community follows an implicit set of rules which may be thought of as values. When these rules are carried into practice, science develops; when they are not, science decays. And the behavior of scientists with respect to their values is sometimes supported and reinforced by society, sometimes frowned on, sometimes actively checked.

One basic value of the scientific community is *freedom of inquiry*. If means of inquiry are ruled out by society, whole areas of science may be suppressed or training in them hindered. Because the sovereign state of New Jersey forbids vivisection of animals, there are almost no medical schools in New Jersey—and this despite the existence in the state of the universities of Princeton and Rutgers. When a means of inquiry is permitted but scientists are not free to follow it as they will, to whatever conclusions they discover, a scientific field is truncated or distorted. This restraint on inquiry usually results from the demand that scientific conclusions be consistent with extrascientific dogma. Such a demand was made by the Church in the case of Galileo. The Nazis would not permit anthropological conclusions inconsistent with

[7] On these subjects, see Bernard Barber, *Science and the Social Order*, The Free Press, Glencoe, Ill., 1952.

their race theory, and the Soviet Union at least frowns upon psychological conclusions inconsistent with Pavlov's behaviorism.

Another important value of scientific activity is *freedom of expression.* The results of free inquiry must be expressed in order that previous mistakes be corrected and a new foundation for further investigation be provided. Today some scientific information falls under security classifications and neither the general public nor scientists who are foreign nationals or thought to be security risks have access to it.[8] Restriction of information, however necessary for extrascientific reasons, hampers scientific progress by limiting the number of qualified investigators who may work on a problem in full knowledge of previous discoveries. Carried further, restriction of expression and information may deprive us of any important contribution to a field.

A third major value is *acceptance and rejection of scientific work entirely in accord with its scientific merit.* It is irrelevant whether the work be done by a moral or gracious man or woman, by a scientist of wealth or social standing, or by a member of an approved race or religion. The Nazis almost eliminated Jews from scientific pursuits and the Russians have limited the number of scientists who come of bourgeois families. The Nazis also opposed the conclusions reached by Jewish scientists of all nationalities and characterized theoretical physics as "Jewish science" because of the part played in its development by Einstein and other Jews.

Another set of conditions favorable or unfavorable to scientific advance is inherent in the social structure itself. Highly specialized *division of labor* makes time available for scientific work. Until the late nineteenth century, universities, industry, and government did not provide jobs for scientists as such. Men without private incomes had to earn their living at other occupations and pursue science only in their leisure time.

Societies with *social mobility* provide every generation with a new elite recruited from a large number of aspirants. Science, like other activities, benefits from the size of the group from which the elite comes. The greater and more equal the opportunities for education and personal advancement, the better the chances for the new elite to be made up of the most talented people. No society has ever offered

[8] It may seem unimportant whether laymen have access to scientific information, but in fact amateurs have made important scientific discoveries.

its members as much and as equal access to education as the United States does today. Of course, other factors may interfere with the process by which an elite is selected—poor quality of education and lack of motivation, for example. Or motivation to enter fields other than science may be so great as to divert many of the most talented. But everything else being equal, the amount and equality of educational opportunity may make the difference between a rapidly developing and a sluggish science.

Despite mass education and great mobility, Americans cannot justifiably congratulate themselves on the percentage of the most gifted people who enter the elite. Recent studies [9] show that less than half of those in the upper 10 per cent of high school classes enter college, and that many do not even complete high school. Of the talented group that goes to college, only about one-half take the bachelor's degree. Poverty is one factor in this waste of human resources. Although it is not too difficult for a student to earn his way through college—and in some schools it is even the rule rather than the exception—there are many cases in which a full-time income from the child is needed to help the family. But poverty is less important as a cause in the United States today than it has ever been. Most schools in the United States, especially elementary and high schools, function at the level of the average student. Gifted children, who are frequently bored and restless when subjected to study far below their capacity, are therefore not motivated to continue their studies in college. In addition, there is a real failure in overcrowded classrooms simply to identify the really promising student. Finally, many families, especially poor ones, are unaware of the opportunities provided by education and assume it natural that their children will not go to college. Partly as a result of this, the National Science Foundation finds a critical shortage today of Americans highly trained in mathematics and science.[10]

The most powerful force unfavorable to science in our time is the highly centralized national state which promulgates doctrines that can be established only by science. The most publicized doctrines of this sort have been Nazi race theory and Soviet genetics. In the in-

[9] See Dael Wolfle, *American Resources of Specialized Talent,* Harper, 1954.
[10] In addition, the methods of teaching and the stimulation of ambition of course play a great role.

terest, presumably, of furthering the belief that human nature is capable of almost infinite modification by the social environment, the Soviet government officially endorsed the theory of inheritance of acquired characteristics—although it has been rejected by biology almost since Darwin. Further, they denied the Mendelian laws of heredity. As a result, Soviet geneticists were purged and Soviet genetics became apologetics, not science.[11]

But one advantage of truth over error is that it may be rediscovered even when expunged from the memory of man. And another advantage is that, when applied properly, it works, which error does not. The Nazis bled Germany of many of her most talented citizens who were "non-Aryans." The Soviet Union deprived herself of the services of middle-class offspring, and expended the time and talent of some of her best scientists in futile attempts to develop new fruits and vegetables by erroneous methods. Soviet genetics may return fully to scientific practice. In competition with genuine science, a spurious politicized science must fall behind.

It is necessary to make three points about the kind of effect society can have on science, so that what follows will not be misunderstood. First, we must always distinguish between the *direction* taken by scientific research under social influence and the *result* of that research, which is not influenced by society. Even if a scientist is hired to do some specific research for government or industry, his conclusions are entirely the result of his scientific work; if they are not, he is not engaged in science. Second, necessity alone is not the parent of invention or of discovery; if one wants to keep the old metaphor, it can be the mother, but only if there is a father as well. Science must be sufficiently advanced; if it is not, no matter how badly an invention or discovery is needed, it will not be made. Society can stimulate work for which science is prepared, but no other kind. At best, society can hasten the preparation. Finally, science, like art and religion, has an internal development of its own. Quite apart from social influence, which fosters some developments and hinders others, science develops its special interests. Even within areas which have been initiated or

[11] Since Stalin's death Soviet genetics has vacillated between science and politics.

advanced by nonscientific pressures, science generates its own problems and proceeds with their solution.[12]

The scientific ferment of the sixteenth century and the rapid development of science in the seventeenth century were not the results of a clean break with all that had gone before and a determination to make science what it is today. The influence of Plato on all Renaissance life—and, through Plato, the influence of Pythagoras, the mathematical philosopher of ancient Greece—led scientists to believe that nature could be explained mathematically, that it was as simple as the principles of Euclidean geometry. Copernicus himself, whose name has been given to perhaps the greatest of scientific revolutions, the discovery that the earth goes around the sun, was inspired by his reading in the classics. His own statement was "I found first in Cicero that Hiketas of Syracuse believed the earth moved." Copernicus was sure that the earth moved, rather than the sun, because he believed that immobility is nobler than movement and the sun is the nobler of the two bodies. Kepler, who discovered the laws of planetary motion, was disturbed that they did not move in circles, for the circle was the most perfect of shapes, but he was gratified to find that the orbit of Mars was an ellipse, because the ellipse was the simplest of all known oval curves. Nature was living up to Plato's idea of her! Galileo regarded observation as subordinate to mathematics as an instrument of science, and Isaac Newton wrote, "Nature does nothing in vain, and more is in vain when less will serve; for Nature is pleased with simplicity, and affects not the pomp of superfluous causes."

Coupled with the influence of the mathematical view of nature derived from Greek philosophy was the Christian doctrine that God's work was perfect, so that no imperfection could be found in nature. Many astronomers and physicists were inspired by the thought that they were proving to all doubters the perfection and the power of God. Newton insisted that the path of the planets could not possibly be ex-

[12] The confusion—an instance of the genetic fallacy—between the social conditions which lead to specific research and the scientific conclusions of that research is common to philosophies which treat some one aspect of society as *the* determining condition for the rest of society. Of these philosophies, Marxism, which regards the economic as the determinant of the rest of society, is the most influential today.

plained by natural causes, and he said in a letter, "When I wrote the third book of the *Principia,* I paid special attention to those principles which could prove to intellectual people the existence of Divine power."

Art, too, has exerted an influence on science. The obvious examples are Leonardo da Vinci and Albrecht Dürer, who were deeply interested in the structure and movement of the human body and in the laws of perspective; their research contributed to anatomy and optics and stimulated scientific work in both fields. And there has been continued influence; the concern of painters with perspective has not ceased and their interest in color and color vision has grown. The French Impressionists of the last century were anxious to learn from science, and their own work in turn led to scientific developments in the physiology and psychology of perception.

More pervasive than any of the matters just mentioned is the way in which science, like all thought, was colored, even dominated, in every age by the basic metaphors which expressed the underlying philosophy of the time. One such metaphor has already been suggested by what we have said of seventeenth-century science: the metaphor of nature as machine. No doubt the metaphor betrays in part an influence of early technology that accorded with other preconceptions, such as that God's perfection was apparent in nature. And this was as important a technological influence as any that will be mentioned later. But it ceased very quickly to have the character of a mechanic's analogy—if it ever was that—and became a first principle of philosophic thought about nature. The mechanical clock has been cited as the most characteristic invention of medieval technology, and it is the clock that became one of the favorite analogies for the "world machine."

The metaphor of the machine suits a nation as technological in its outlook as the United States almost too well, for we will not drop it. In the 1920's John B. Watson, pioneer behaviorist in American psychology, popularized the metaphor as it applied to man, and lecture platforms were the scenes of interminable debates on the subject: "Is Man a Machine?" Watson and his followers were deluded by the analogy of the machine into a reductionist fallacy. To prove that man was, or was like, a machine, they had to deny the distinctive character of thought, learning, memory, imagination, and purpose. All these they reduced to the pattern of stimulus and response, a pattern they regarded as mechanical.

But the same metaphor that misled Watson is of help to scientists today. Physiology and psychology have advanced, increasing our knowledge of man, and electronics has changed our concept of machines. Cybernetics, the study of both the human mind and machines, has been created. The question at the moment is not "Is man a machine?" but "Can machines think?" Instead of denying the character of mental functions, contemporary scientists are constructing machines that can duplicate mental functions. Then, through the study of the new machines, it is hoped that there will come greater understanding of the mind.

Politics and law have contributed more than their share of metaphors to science. It is easy enough to see how political and legal ideas would seem to be statements about the order of nature. There is, in the first place, the ambiguity of the word "law," which we pointed out in discussing misinterpretations of science. In the second place, there is a tendency to regard the political and legal customs of one's own nation, especially if they are fervently believed, as natural and proper for all nations. We have mentioned the "body politic" as a political metaphor derived from biology; yet there is a long history of biological metaphors derived from the political state.

Alcmaeon of Croton, a Greek of the sixth century B.C., described physical health as a balance of the qualities that made up the body. This state of equilibrium was called *isonomia* in Greek, and also meant equality of political rights. So Alcmaeon wrote of disease as imbalance, or domination by one of the qualities, and called it monarchy. Galen, court physician to the emperor Marcus Aurelius, used a Roman legal concept of justice—"to each his own," or everyone sharing in accordance with rank and status—as a basic metaphor. Some parts of the body, he said, have few nerves; this is an example of justice because they do not need much sensitivity.

To move closer to our own day, Rudolf Virchow (1821-1902) emphasized the cell as the unit of life. What, then, of the whole body, the organism itself? That, said Virchow, ". . . amounts to some kind of social institution." What kind? Virchow was a democrat and a liberal of a dedicated sort, and his answer was, "It is a free state of individuals with equal rights though not with equal endowments which keeps together because the individuals are dependent upon one another. . . ." Not so far from Alcmaeon for all the intervening centuries! [13]

Let us conclude our discussion of metaphors in science with a quotation from Carl Becker.[14]

> If we would discover the little backstairs door that for any age serves as the secret entranceway to knowledge, we will do well to look for certain unobtrusive words with uncertain meanings that are permitted to slip off the tongue or the pen without fear and without research; words which, having from constant repetition lost their metaphorical significance, are unconsciously mistaken for objective realities. In the thirteenth century the key words would no doubt be God, sin, grace, salvation, heaven, and the like; in the nineteenth century, matter, fact, matter-of-fact, evolution, progress; in the twentieth century, relativity, process, adjustment, function, complex. In the eighteenth century the words without which no enlightened person could reach a restful conclusion were nature, natural law, first cause, reason, sentiment, humanity, perfectibility (these last three being necessary only for the more tender-minded, perhaps).

Are there any intellectual influences on social science especially? Of course. Social science, which deals with matters so much closer to the daily concerns of the culture in which it exists, is likely, we have seen, to be more influenced by culture patterns than natural science is. One influence that has left an enduring mark is that of natural science. Social science, like any younger brother, is deeply affected by the accomplishments of the first-born, and in this case is moved to emulation. Not only the success but the prestige of natural science is a source of envy to the social scientist. So he borrows concepts—the metaphor of the machine, evolution, the action of forces—and imitates procedure. He learns to use mathematics, sometimes where it is of no avail, and he develops laboratory and experimental techniques.

In addition, as social science grows it shows the effects of national environments. American sociologists, for example, do much more work than their English and European colleagues in race and

[13] For those who want to pursue the subject, see "Metaphors of Human Biology," by Owsei Temkin, Professor of the History of Medicine at Johns Hopkins University, which appears in *Science and Civilization,* Univ. of Wisconsin Press, 1949.

[14] *The Heavenly City of the Eighteenth-Century Philosophers,* Yale Univ. Press, 1951, p. 47.

minority group relations, subjects especially pressing in the United States. American social science also reveals something of American attitudes in being preponderantly empirical. It is true even in natural science that Americans tend to contribute to empirical research and Europeans to theoretical. The close relationship in America between physics and engineering, chemistry and agriculture, is still unique, but large-scale theoretical work comes for the most part from Europe or from Europeans in America. The difference is probably more marked in social science. A typical statement urging the purely empirical is this one, made in 1949 by Professor William F. Ogburn, one of the founders of American sociology: [15]

> The scholarly tradition is an obstacle, in my judgment, to social science as it was to natural science. For the literary scholarly tradition values verification somewhat less than does the tradition of science and gives more acclaim to other intellectual virtues such as analysis, interpretation, argument, discussion, and imagination. The literary scholarly tradition often accords high honors to publications displaying these virtues even when the question "how do you know it?" has not been answered by evidence. Of particularly high prestige, unfortunately, in the social field is synthesis, or the construction of systems of ideas. Thus in economics the person who develops a theory—that is, builds a system of ideas—often acquires more honor than one who verifies a hypothesis. In other words, prospective social scientists do not hold steadfastly enough to the discipline called for in verification. Instead they are drawn off into intellectual display where the rewards are still very great.

Although American social science was already highly empirical, and many extremists disdained theory, Ogburn wanted to go still farther, and misrepresented the situation in his zeal. The reverse was probably the case.

For good or ill, the rate of social change is constantly being accelerated. Every decade seems to contain more changes than the half century before it and these changes are, to an enormous extent, the results of modern invention: the electric light, the telephone, wireless, radio, television, propeller-driven aircraft, jet aircraft, atomic energy. Of course, to get these, society had to approve research, and to use them, it had to be receptive to technological innovation. But granted

[15] *Science and Civilization*, Univ. of Wisconsin Press, 1949, pp. 205-06.

these conditions, which surely characterize our society, technology has a pervasive influence on our lives.

Although technology is the most obvious social consequence of contemporary science, it is not the only one. Acceptance of science fosters a pragmatic and engineering attitude which leads to the manipulation of the environment, physical and social, for the deliberate attainment of ends. This attitude has probably weakened religious belief and the acceptance of many traditional values. It has also led to a cult of the pseudoscientific, which nurtures irrationality.

Technology has become part of us in a more intimate way than the scientific outlook has. Life is much more comfortable, and physical mobility is increasing with the greatest rapidity. Within the last hundred and fifty years there has been more technological change in Europe and America than there was for a thousand years before. The chief form of conveyance at the beginning of the nineteenth century was the horse or horse-drawn carriage, exactly as it was in ancient Rome or, for that matter, among the Mongolian hordes of the steppes. Agriculture was conducted essentially as it had been on the farms of the Middle Ages. Oil and wax still lit houses at night. Communication was so slow that Andrew Jackson and his troops fought the battle of New Orleans months after the War of 1812 had been officially ended by treaty. Roads and dams were not nearly so good as those the Romans used to construct; neither was plumbing.

In these hundred and fifty years applied science in the form of medicine, higher standards of living, and sanitation has been a factor in increasing population and extending the individual life span. Population at the beginning of the Christian Era is estimated at two hundred million people in the entire world. Population of the world had perhaps doubled by the year 1776. It grew perhaps 75 per cent again by 1860. In the last one hundred years population has increased about 150 per cent and it is growing so rapidly that in the next hundred years or so it may increase 600 per cent. Today there are almost three billion people in the world. P. K. Whelpton, former director of the U. N. Population Division, said that if population had grown from the time of Christ as it has in the last one hundred years there would be, for every human living today, one million more. New York City would have, not eight million people, but eight trillion, i.e., three thousand times the earth's population today.

Surely this immense increase has contributed to the intensity of pressures felt, especially by the urban population, in our day.

Not only are there more of us, but we are older. On the average, the life span in the United States is seventy years; in 1800 it was about thirty-five years. Countries that are technologically backward, especially crowded ones, are far behind our standards of longevity, even behind our standards as of 1800. Life expectancy in India, for example, is little more than twenty-eight years. A tremendous infant mortality accounts largely for this low figure, but even when an Indian survives infancy, his life span is less than that of Americans taken as a whole. Further, women outlive men by as much as five years in all European and American countries, but Indian women do not live as long as Indian men.

Law, industry, and social mores are affected by the increased average age of our population.[16] Either people must work to a later age than before or save more money while they do work, or they become a financial burden for many years on their children—and, through pensions and social security, on business and government. The tax structure can be revised to meet this new expenditure, and businesses can change their retirement rules, but there is another problem which cannot be met so directly: What are retired people to do with their leisure time? This is part of the general problem of increased leisure brought about for all age groups by more money and fewer working hours. The entertainment industries have boomed in consequence; Florida, California, and Arizona have new settlers of advanced years; and adult education has become a major part of American education.

What is still not clear is the effect of this ever-growing age group on political attitudes and social customs. Obviously, there are new interests to be served by social security plans, and so there is some movement toward a welfare state. But what about public attitudes toward war and peace, freedom, initiative, personal security, love and marriage, when the public is composed of age groups in a different

[16] Most people start work between 15 and 25 years of age. Yet that age group is a diminishing percentage of the adult population. So, although there is a trend to earlier retirement, the average age of workers is rising. In 1961, one in every eleven Americans was 65 or more, compared with one in twenty in 1900. Most startling, the growth in the group over 85 has been 920 per cent since 1920.

numerical proportion than ever before? Is such a public more "liberal" or more "conservative"? Is it partly responsible for the changes in the political climate of the last decade? These are important questions for the social sciences. When we have some answers to particular questions like these, we may also have the outline of an ongoing trend.

Technology is also responsible for an enormous physical mobility and for an increase in the social mobility inherent in our system. Although the airplane is the most dramatic symbol of the revolution in transportation, it is the automobile that has so far affected our lives most. Sociologists and novelists have given us a picture of the alteration in family life brought about by the automobile: the drives on week ends, the dating and courtship patterns of adolescents, the roadside motel. We are a nation on wheels and we move about restlessly.

Physical and social mobility have, to some extent, grown together. As it gets easier for labor to move and be moved, our economy grows more flexible and opportunities for advancement that depend on physical movement are more readily seized. Innovation, especially in the form of laborsaving devices, releases workers from one job and creates a demand for them in others which may be geographically distant. To move to a new locale in order to profit economically is also to give up friends and to break the ties of community that are often the source of emotional stability and happiness.

The premium placed on innovation raises the level of aspiration and this in turn yields more innovation. To many Americans there seems no end to this process; and since the prospect of constant technological advance fits so well with other beliefs—the optimism of democratic and egalitarian attitudes and a religious notion of the perfectibility of man—it bolsters a basic social myth, progress.

For the moment, the belief in progress is countered by fear of world destruction. We seem to alternate between dread of a future shattered by deadly atomic weapons and hope of a future which combines H. G. Wells with Buck Rogers: prosperity for everybody, almost all work done by the machine, spaceships traveling on schedule through the planetary system, a world which is really a vision of New York City developed in mechanization and made universal.

Yet the greater dangers are probably not in the prospects of destruction but in the future we hope for, and many of them already exist. Many traditional values of American life—self-improvement, advancement by merit, the possibility that every man will reach the top, novelty, adventure, and excitement—can be realized today in terms of technological advance as they were once in terms of the frontier. But there is now an enormous price to pay in insecurity and anxiety. The greater the rate of change, the more difficult it is to adjust to the new situation; and the more we emphasize the future, the less we remember the past. If every day brings something new, yesterday was little preparation for it. Competition may yield progress, but it also yields insecurity and the fear of failure, and novelty and adventure are not easy to reconcile with stability and peace. An everchanging and exciting society is likely to be full of restless people, living frantically in a pursuit of success and approval, having no time for contentment, regarding serenity as an antique or medieval ideal.

A peculiar tragedy of our time is the split between life and work, a division made more extreme by the machine. Tending the machine for seven or eight hours a day is the lot of many of us,[17] and it is work that is properly called "mechanical," work in which we ourselves behave as though we were machines, and in which little is asked of us except a monotonous, machinelike accuracy. It is an old story that the age of the craftsman is behind us. On the modern assembly line, men are not able to carry through the making of a product from its inception to its completion, garnering the satisfaction of creativity. Rather, they work constantly at a single operation, usually not knowing what precedes or succeeds it. At its worst, the task of the machine tender is the task of the robot; and only when the worker of the machine returns to his home after his day's labor does he begin his real life. Creativity, which was once the birthright of every craftsman, is now limited to the designer in the drafting room. A life in which basic interests extend to work is usually led only by the artist, the scientist, the entrepreneur, sometimes the physician, lawyer, and teacher. The machine has liberated us from actual slavery; it is one of our greatest problems to see that it does not, like slavery, make us subhuman.

The scientific outlook, both of method and conclusion, has become part of our image of ourselves and our place in the world. Looking first at the effect of the conclusions of science, we find that chemistry, physics, and physiology have helped create the belief that

[17] Not only of "workers," but of the white-collar tenders of typewriters, calculators, check-writing machines, and so forth.

man is a machine, more complicated perhaps than the machines he creates, but of the same general type. Astronomy and biology have yielded new perspectives on man's place in nature; the newly discovered vastness of the physical universe leads man to see himself as a parasite clinging to a tiny ball in a corner of one of the smaller galactic systems.[18]

It is a commonplace that the medieval view was that this is an earth-centered universe in which all physical things were made for man. But it is worth quoting Anatole France's description of that world in *The Garden of Epicurus:*

We have some trouble in picturing the state of mind of a man of olden times who firmly believed that the earth was the center of the world and that all the stars turned round it. He felt under his feet the souls of the damned writhing in flames, and perhaps he had seen with his own eyes and smelled with his own nostrils the sulphurous fumes of Hell escaping from some fissures in the rocks. Lifting his head he contemplated the twelve spheres, that of the elements, containing the air and fire, then the spheres of the Moon, of Mercury, of Venus, which Dante visited on Good Friday of the year 1300, then those of the Sun, of Mars, of Jupiter, and of Saturn, then the incorruptible firmament from which the stars were hung like lamps. Beyond, his mind's eye discerned the Ninth Heaven to which saints were rapt, the Primum Mobile or Crystalline, and finally the Empyrean, abode of the blessed, toward which, he firmly hoped, after his death two angels robed in white would bear away, as it were a little child, his soul washed in baptism and perfumed with the oil of the last sacraments. In those days God had no other children than men, and all his creation was ordered in a fashion at once childlike and poetic, like an immense cathedral. Thus imagined, the universe was so simple that it was represented in its entirety with its true shape and motions in certain great painted clocks run by machinery.

We are done with the twelve heavens and the planets under which men were born lucky or unlucky, jovial or saturnine. The solid vault of the firmament is shattered. Our eye and our thought plunge into the infinite abysses of heaven. Beyond the planets we discover no longer the Empyrean of the elect and the angels, but a hundred million rolling suns, escorted by their cortege of obscure satellites invisible to us. In the midst

[18] The effect of this perspective on older religious views of man is dramatized by Wilbur Daniel Steele in a short story, "The Man Who Saw Through Heaven." See *The Best Stories of Wilbur Daniel Steele,* Doubleday, 1946.

of this infinity of worlds our own sun is but a bubble of gas and our earth but a drop of mud. . . .

What is seldom perceived about the medieval world view is that because it was God who had created and sustained the universe and man in this fashion, man—though the center of the *physical* universe —was weak and dependent in his more important capacity as a *spiritual* being. The universe was physically man-centered because it was spiritually God-centered. But modern science removes man from the center only as a physical creature in a physical world. Even if no place is found for the soul, the intelligence, which discovered and formulated the laws of science, takes credit for having worked the transformation by which man was removed from the center of things. We may be tiny, unimportant creatures, but it is we who know it and we who say so. The universe dwarfs man's body but is plumbed from end to end by man's mind. Indeed, we will call a statement true only if it can be verified by man's methods and senses. Once again, as for the ancient Greek Sophists, "man is the measure of all things." He has taken himself from the physical center of the universe, where medieval man had placed him, and put himself in the spiritual center, where medieval man placed God.

Instead of humility resulting from this change in physical conception, as many scientists expected, man became arrogant. As some of Dostoevski's characters say, if there is no God, then man is God; or as Nietzsche's Zarathustra puts it, since God is dead we must prepare the way for the Man-God.

The methods of science have perhaps affected our outlook less than the picture of the world implicit in scientific conclusions. To be sure, there may be a more widespread demand today than ever before for evidence of a sort science would accept, and a skepticism about knowledge that is not scientific. But we tend to replace beliefs we have come to doubt with others no less dubious and often much more dangerous. Belief in religion is replaced by religious devotion to political belief, and redemption in heaven by redemption on earth. And we are often naïve when we most want to be "scientific" about the views we have inherited. We think that we are no longer entitled to once cherished beliefs and biases which cannot be supported by scientific information. Yet all that an inability to find scientific warrant for our beliefs means is that they cannot be accepted in the body of science, or that our rationalizations are poor. It does not follow that the beliefs should be abandoned—that religious people, for example, are obligated to become atheists or agnostics, or that we should regard moral virtues like honesty as matters of taste.

Science has perhaps had its greatest effect on man's outlook indirectly, through technology. This is an age of machines and gadgets, and they stand between us and the simple essentials of daily life. Between the housewife and the dirty floors, dirty dishes, and dirty clothes are the vacuum cleaner, the dishwasher, and the automatic washing machine. For a long time now, the abattoir, the butcher, and the grocer have stood between the city dweller and the procurement of food, but today the picture magazine, the comic strip, and the digest have come between man and his books. The enormous rate of literacy—but literacy without real education—yields a mass market for the great industries of "popular culture" that create our entertainment in patterns as rigid as if they were stamped out by machine.

The social scientist, too, is in a new situation. Although exchange of information with his colleagues is not limited so much for the social scientist as for the natural scientist, some social research may be classified as secret: morale problems of soldiers under battle conditions, responses of civilian populations to saturation bombing, and so on. But the major problem is an atmosphere of some hostility to public discussion of controversial social issues. The social scientist must ask himself what role he should play in society, what sort of influence he wants to have, and how he can have that influence under contemporary conditions.[19]

[19] See Gunnar Myrdal, "The Relation Between Social Theory and Social Policy," *British Journal of Sociology,* Vol. IV, No. 3, September 1953, pp. 217-22.

# 7

# Science, Morals, and Values

Before reflection we identify moral behavior with conventional behavior, and convention or custom with what is morally right. In isolated, *immobile* societies custom changes little, if at all, over the centuries. Although a man or woman may have problems of moral behavior in those societies (when, for example, desire conflicts with custom), he rarely has problems of moral belief; he believes what everyone else in his group believes.

Logically, it is possible at any time to ask of any conventional act, "I know this is customary, but is it right?" Psychologically, it is most unlikely that such a question will be asked until a difficulty is felt, and difficulties about morals arise most when there is cultural contrast or change. Cultural contrast comes when acquaintance with another culture—through war, travel, or commerce—brings awareness of a different set of customs, and questions arise like "Whose customs are right, theirs or ours?" or "Is it true that their customs are right for them but our customs are right for us?" Change may come for a number of reasons: cultural contrast itself; borrowing from other cultures; new ways imposed after defeat in war; new geographical environment due to climatic shift or migration. When change occurs in any part of the culture, there may be inconsistency with the older parts; abundance of game in a new locality may make it seem absurd to have days of religious ritual before a hunt. And change may render old customs irrelevant or actually opposed to survival or efficiency.

When Jerusalem was besieged by the Roman armies of Pompey the Great, the Jews defended the city successfully for a time, but when a spy told Pompey to attack on the Sabbath, he found an unresisting population immersed in prayer.

Not only does change raise questions of moral belief; it also creates conflict. If a Jewish soldier had urged that his people use arms on the Sabbath for the duration of a war, he might have been denounced as impious. *Fiat iustitia, pereat mundus* ("Let justice be done though the world perish"), said an old Roman proverb. The presuppositions, of course, are that we know what is just or right; it is ancient custom, and it is not right simply because it is custom, but it is customary because it is right. This last presupposition is the rationale behind most unwillingness to change moral belief in the face of changed conditions of life. Few people defend custom sheerly as custom. They assume it is natural, *the* way for men to behave; or God-given, obedience to divine command.

What we call Western civilization developed through migrations, wars, and trade, and it has been in constant change. So moral reflection in the West has a long history. But as the rate of social change increases—and modern technology accelerates it tremendously—customs, too, change more rapidly and problems of moral belief become more urgent. After all, moral principles, explicit and implicit, guide our decisions and actions; they are the ultimate justification for what we do. When we enter new situations in which our former beliefs are inadequate, we must either change those beliefs, fail in our actions, or become hypocrites, giving lip service to the old beliefs but acting in accordance with new ones which we do not acknowledge publicly and perhaps never formulate even to ourselves.

It has been claimed—and denied—that moral belief itself should be changed because of the conclusions and the intellectual methods of science. Two major sets of questions arise. (1) Can science determine whether it is possible for man to act morally? After all, morality exists only when there is choice; if we *must* do something there is no point in asking whether we *ought* to. And science presumably casts light on the extent to which choice is free or conditioned. Has physics established that man's will is free? Has psychology refuted free will? (2) Can we apply scientific method to verify or refute moral principles? But first, what is the nature of moral belief and what kinds of problems are moral problems?

We imbibe thousands of moral notions with our mother's milk, but that is no justification for them. Society may accept some action as right, but we may still ask, "Is it?" Why is it right to have only one wife and to cherish one's parents? Why is it wrong to murder?

Social scientists sometimes try to eliminate these questions by pointing out that each society conditions its members to think that the moral beliefs of their culture are correct, and that other cultures have their own beliefs which are accepted with equal devotion. But that merely avoids the questions; it does not eliminate them. Of course people are conditioned to believe one thing or another and each culture has its own beliefs. But are they the right beliefs? Ought they to have them? Put another way, are there good reasons for believing that these beliefs are right? That some people believe in cannibalism scarcely impels us to accept it as right, even for them.

But we ourselves, who ask these questions, are conditioned to accept some moral beliefs and to reject others. Can we really question them? The fact that we do is proof that we can. The chief intellectual danger of our conditioning is that when we try to give answers we may do nothing but provide apologies for what we believed in the first place. But with care even that danger can be avoided or minimized. If, then, we pursue the question why we ought to accept a particular belief, we soon find that any answer can itself be questioned. "You ought to exercise." "Why?" "To be healthy." "Why be healthy?" "To live long." "Why?" And we can keep asking, "Why?" until we get to what seems a final answer: "Unless people accept this as right there will be no society at all." But probably we can still ask meaningfully, "Why should there be society?" What most people who ask such questions seem to be looking for is a final answer that itself cannot be questioned.

Such an answer would be categorical and would have the form, "One ought to do *A* because of *B*." The reason that the answers so far considered seem unsatisfactory is that, on analysis, they turn out to be hypothetical, having the form, "*If* you want *B*, *then* you ought to do *A*." [1] If you want to live in society then you ought not to murder; if you want democracy to survive you ought to allow free speech. But what if someone does not care about society or democracy, or is

[1] Immanuel Kant argued that a true moral rule must be categorical, not hypothetical. Hence he spoke of the basic moral law as a categorical imperative, or command.

actually opposed to them? Then the rule, or belief, does not seem to apply to him. In the hypothetical form, a moral rule is only a means to an end. When we say, "If you want *B*, then you ought to do *A*," we can still be asked, "Why?" The answer is that if you do *A*, then you will attain *B*, and it is assumed that *wanting* an end—not merely day-dreaming about it—implies wanting the means to attain it.

Some people are willing to rest content with only hypothetical moral statements, but most people feel the need of categorical statements. They will not accept the idea that what one ought to do depends on the ends he has chosen, but insist that morals should specify what ends a man ought to choose. So instead of the statement "If you want *A*, then you ought to do *B*," they would like the statement "You ought to want *B*, and therefore you ought to do *A*."

But how can this be a final answer, precluding the question why one ought to want *B*? The only way it could be final, making it absurd to ask, "Why?" would be by asserting that it is somehow in the nature of things that men ought to want *B*. Specifically, there are three important ways in which this has been said, and each is regarded by its adherents as the proper base of moral thought. A thing is right because (1) God wills it, (2) because Nature made it so, or (3) because the nature of man or the structure of his mind makes it necessary. So if we say that a thing is right because God wills it, it is not reasonable to ask why God wills it; we define God so that it is His nature to will it, and God's nature is, by further definition, infinite and beyond our finite understanding. And if we say that it is Nature that makes a thing right, it is silly to ask why Nature is as it is.[2]

The difficulty with these answers is that although we can no longer ask, "Why?" we may readily ask, "How?" How do we know what God wills or what Nature means for morality? And there is a different form of the same question. Can we reason logically from whatever we know of Nature or God or man to the principles of morality?

First, how do we know what moral behavior God wills? There are many people who will gladly tell us, but how do *they* know? Some of them claim to have it directly from God Himself, or at least to have learned it from others to whom God revealed it. But there are conflicting claims to revelation and claims to conflicting revelations. How can

[2] Of course, we might ask why we should obey God or Nature. Perhaps we have no choice. But if so, there is no moral issue.

we choose among them? There is no principle in terms of which we can choose logically. People believe one set of revelations rather than another as a matter of faith, and the vast majority of religious people believe the revelations accepted by the religion in which they were raised. For many purposes this behavior is quite satisfactory but, intellectually, it scarcely provides a logical principle of choice.

An equally difficult problem exists if we choose the second of these moral bases, Nature, or the structure of the universe. Nature has never spoken its will to us. If that seems a naïve objection, is it less naïve to think that moral rules can be discovered by an examination of nature? We cannot tell, no matter how much we know of physics, chemistry, and astronomy—the sciences which constitute our knowledge of nature—what moral rules are consistent with them. And biology might let us know what would keep man healthy, but can we equate organic health with morals? The martyrs of history would toss in their graves at that answer.

Third, suppose we believe that that conduct is right which is natural for the human organism. How do we know what is "natural"? If nobody had ever performed a particular action, and it seemed impossible for man, then that action might be thought of as "unnatural"; it would not, however, constitute a moral problem. No one, by the use of his body alone, ever flew, and there is no moral problem connected with it. Moral problems arise because of things people do. If we could fly, human flight would have to be regulated by law, because there would be special moral problems of flight. People would be able to peer into other people's rooms, they might collide or have other accidents, the crime rate might go up.

And can we reasonably say that anything that men actually do or have done is unnatural? What would "unnatural" mean in that case? Some people have attempted to answer this by examining the animals. Not finding homosexual activities, for example, in other animals, they have argued that it is unnatural in man. More recent study, however, *has* revealed homosexual behavior in other animals—but neither "fact" is germane to the argument as to what is "natural." If it were unnatural for man to do anything that chimpanzees did not do—to take perhaps the most intelligent form of nonhuman life on this planet—then it would be unnatural to write poetry, play chess, or even examine other animals to discover whether their behavior is like man's.

As for the notion that moral principles result from the character of the human mind, it may be that the idea of morality depends on the structure of the mind. But no *particular* moral principles can be derived from that fact, otherwise we would not encounter the moral differences and disputes which exist between cultures and within cultures. Men, having the same kind of minds, would have the same moral principles, and knowing what was right and good would be as easy and natural as breathing.

The belief in either God or Nature as an absolute moral base probably yields the same psychological satisfaction and the same assurance of being right. Some thinkers, like Thomas Aquinas, even combined the two.[3] But a line has often been drawn between the adherents of God's will as the basis of morality and the adherents of Nature. The former are said to rely on religion or faith to reveal moral truth, the latter on reason—although reason cannot prove that Nature is the base of morals. This ancient dispute reappears occasionally in our day as a quarrel between religion and science.

Here the effect of social science is more direct than the effect of natural science. Natural science has no moral content; social science includes *descriptions* of moral belief and moral behavior, and may legitimately attempt to formulate laws stating the social conditions under which one or another set of morals is accepted, rejected, or modified. But social science does not include *prescriptions* for moral behavior; it does not evaluate belief in moral terms.

Although social science does not urge any moral belief or action, it provides information that can be used in moral appraisal. If we learn, for example, that the consequences of acting on a particular moral belief under specified circumstances are destructive of another moral belief, we know that under those conditions we must choose between the two beliefs, that we cannot successfully apply both. So we may believe that any person is entitled to increase his wealth and we may also believe that individual wealth should not be increased at the expense of others. Under particular economic conditions, we may act on the first belief without violating the second, but there are other economic conditions, we learn, under which the two are not compatible.

[3] Thomas believed that God's Eternal Law, which guided the universe, was basic, and that Natural Law was derived from it. Human Law, he thought, should be based on Natural Law, which could be discovered by reason alone. Divine Law, which led to salvation, came from revelation.

Yet the practice of science, as distinct from its content, does presuppose, we have seen, some moral beliefs on the part of the scientist. He must accept truth as the fundamental value in his work. This bars him from making scientific assertions that he regards as false, even if he thinks it good for people to believe them. It should also make him willing to listen to and examine any scientific proposal on its merits, for truth may come from any source. These commitments may bring the scientist and others who share his values into conflict with religions which proclaim an absolute revealed truth and oppose utterance or publication of contrary belief. But this conflict would be about social and moral matters like freedom of speech and thought, not about any of the specific findings of science or the doctrines of theology. We recognize that quarrels between religion and, for example, a mechanistic view of biology or psychology, which implies that there is no entity such as the soul, are not quarrels with science. Science does not pronounce on such matters—by definition, the soul is immaterial and so beyond physical observation; but these are quarrels with philosophies of science which interpret scientific data in a special way.

The current dispute aired most frequently is about the absoluteness or relativity of morals. Modern Western religions preach an absolute morality, unchanging except in its applications to different times and situations; science is popularly supposed to favor the relativity of morals. Here our distinctions can show this quarrel to be based on misunderstanding. The social sciences discover the relativity of existent, or conventional, moral belief. They do not imply—although some social scientists have mistakenly tried to infer it—that the moral beliefs people *ought* to hold are relative. It is perfectly consistent, though not necessarily true, to believe both (1) that the moral beliefs men actually have are relative to (i.e., dependent on) their culture, and (2) that most of those beliefs are false or immoral because there is a single correct moral standard.

Unless there is choice there is no moral situation. Choice exists when there are alternatives. If there is only one thing to do I do not have choice; if there are alternatives but I am constrained by external power to do only one, I do not have effective choice. In the simplest instance, if there is a fork in the road and I am pushed to the left by someone stronger than I am, I may have had intellectual choice and decided to turn right, but I cannot reasonably be said to have had the effective choice of going either left or right. I was powerless to carry out my decision. Of course, the external power need not be another person. If I fall from a window and land on someone's head, it is not likely to be from choice; where I have no control or power I cannot *act* from choice. And where I have no choice at all, there is no moral issue.

Choice cannot be evaded by unwillingness to consider possible alternatives or by sheer refusal to choose. Such refusal is itself a choice, even when the refusal is only implicit. If I will not choose to go either left or right when the road forks, I have chosen to remain where I am or to retrace my steps. If I witness a fight to the death and refuse to interfere, I have chosen to let murder be done. Pontius Pilate washed his hands, symbolically preserving his integrity, when Jesus was condemned to death, thus implicitly choosing the execution of Jesus rather than the consequences of Roman intervention. Nor can we foist moral responsibility for our choices onto others. We may be members of an organization—church, or political party, for example—that makes decisions for us, but retaining membership, even if we were born to it, is a choice.

Although choice is a necessary condition for every moral situation, it is not a sufficient one. Situations that involve choice may be classified as those of (1) taste, (2) advantage, (3) morals. There are some situations which include two or all three of these elements.

1. Choices of taste are those in which no reasonable argument can be made that one is "better" than another. This has been recognized in many adages, among others *de gustibus non est disputandum* and *chacun à son goût*. If I prefer chocolate ice cream and you vanilla, or I prefer ale and you beer, that is all there is to it. It would be absurd for either of us to argue that the other's taste is wrong or bad.

2. Choices of advantage are those in which an argument can be made that one choice is preferable to another, in the sense of being more advantageous, because it leads to something desired. That is advantageous which is, in the language of ethics, an efficient and economical *means* to a specified *end*. Now, disputes about the choice of means may be quite rational. Which of two golf clubs is preferable for a particular stroke? What course of college study is best for a potential lawyer? Which foods should be eaten by a boxer in training?

These questions may be argued and answered. Note that "means and ends" are equivalent to "causes and effects"; means are causes which yield ends as effects. We use "means and ends" when we talk about purposes and plans; we use "causes and effects" when we talk about natural happenings that are not brought about by an agent, and so involve no reference to purpose.

Some questions that are dismissed as entirely matters of taste are also matters of advantage and of morals and can be discussed with profit. Aesthetic choices, for instance, may include all three categories. If I prefer Shakespeare and you prefer Mickey Spillane, it can most reasonably be argued that, except under circumstances of fatigue, sociological study, and the like, it is better (more advantageous) to read Shakespeare. What is done to the mind, the sensitivity, and the emotions of the reader by each of these writers is very different, as is the quality of the aesthetic experience. And, as a moral matter, what is done to civilization as a consequence of taking one or another seriously is equally different.

The very word "taste," in the sense in which we have been using it, is a metaphor of the palate, of taste in the mouth. It implies immediate sense response. We like some things and dislike others, but not for any reason; we are made or conditioned that way. Yet taste can be deliberately trained; we speak of "acquired tastes," like the enjoyment of olives. And it is obviously a help if our tastes are trained so that we like what is advantageous and moral. Then we can rely on immediate response to guide us in some things and we do not have so often to make ourselves do what is moral although we dislike it.

Taste, then, can be a matter of standards to which we are habituated, a liking for the music of Vivaldi, for instance, or for gentle manners. And the training of taste is basic to education. John Dewey said, "The formation of a cultivated and effectively operative good judgment or taste with respect to what is esthetically admirable, intellectually acceptable and morally approvable is the supreme task set to human beings by the incidents of experience."

Even taste in the sense of immediate enjoyment remains unarguable only when it is isolated from the kinds of social circumstances which convert it into an issue of advantage or morals. If I like chocolate but chocolate is unhealthy for me, it is to my advantage to give it up. If there is a chocolate shortage and I want more than my share, or if chocolate can be used for national defense and I want to flavor ice cream with it, my taste for chocolate has become a moral problem. There are two ways of dealing with these situations. I may regard advantage or morals as irrelevant to my taste for chocolate, but as basic to the decision whether to indulge it. Then, if I ought not to indulge it, I try not to allow my taste to lead to action. Or I may try to change my taste to, say, vanilla, so that I need not feel repressed.

3. Every moral philosophy seems to mean something different by the word "moral." But there is general agreement that "moral" refers to what ought to be, and that moral choice exists when, among the alternatives that confront us, one ought or ought not to be chosen. Some philosophers insist that "ought" has no special moral meaning, that it signifies only the advantageous, whatever is a means to an end, as in the statement "If you want *B*, then you ought to do *A*." But that is not what is meant by the word "ought" as it is ordinarily used in a moral context. When we say that as a moral matter you ought to pay your debts, we do not mean that you should do so for the advantage to your business reputation. If no one could possibly know, it is to your advantage to default, for you save money without injuring your reputation. Still, morally, you ought to pay; there is an obligation to do so.

A moral situation, then, is a situation containing choice in which the alternatives relate to obligation, not just to taste or advantage. Every society treats some actions and some restraints on action as obligatory, and demands that its members do the one and not the other, whatever their tastes or advantage. The concept of the obligatory is a difficult one to define, but it includes demands made on conduct, perhaps expectations that people will act in one way rather than another, and it takes precedence over taste and advantage. An equivalent, perhaps identical, concept is that of duty. It is often argued that in some larger sense it is more advantageous to be moral, to do one's duty, fulfill his obligation, than to seek specific momentary advantages. This may indeed be so, but it is confusing to use the word "advantage" in two senses at once, and moral behavior often does lead to the advantage of others or of society while at the same time yielding personal disadvantage. To support a widowed mother may be obligatory, but it costs money which might be expended on more direct satisfactions.

The questions we asked about the base of morals apply fundamentally to obligation, or duty; they are not relevant to problems of taste or advantage. When we learn, for example, the conventional obligations imposed on, or accepted by, the members of any society or any group, we can still ask whether they ought to be the obligations of those people. This would involve us in a distinction between conventional morality and true, or critical, or reflective morality. And we would have to find out how a critical morality is arrived at. But that is beyond the scope of this book. What we have been trying to do is to distinguish the specifically moral from taste and advantage, and the "ought" of rational calculation of advantage from the moral "ought" of obligation; to show how conflicts between the individual person and his society can be conceived in these terms; and to clarify the role of science and reason in matters of taste, advantage, and morals.

What men prize, cherish, or desire is ordinarily called their values. And one may contrast the valued with the valuable, which is what men *ought* to prize, cherish, or desire. Now "valuable" may mean what ought to be valued either in the sense of what leads economically and efficiently to a specified end or in the sense of moral obligation. But whichever meaning we intend, we may criticize actual values by arguing that they are not valuable.

Values are probably best thought of not as particular ends that are desired, but as criteria or standards in terms of which we choose or evaluate such ends. Justice, liberty, and democracy may be such standards, as may wealth, self-fulfillment, and happiness. These are not things or situations, and so cannot be directly *attained;* they are *realized* only by the attainment of ends which yield them. So when I choose an end, a desired outcome of action, like increasing my salary or marrying the girl I love, and I act in such a way (means) as to attain them, I am trying to realize values, probably wealth and happiness. We realize a value when we attain ends that conform to that value as a criterion.

Reason is relevant to values in two important ways. The first is logical: values should be consistent with each other. The second is scientific: the means we employ should in fact bring about the ends we desire. If our values are inconsistent, realizing any one of them may destroy the possibility of realizing others. Except for an extreme—probably nonexistent—fanatic, no man has only one value, and it is even rare that a man has a single value so much more important to him than all others that he would sacrifice all the rest for that one.

Most people, if they analyze the values they hold, will discover great inconsistencies. Little thought is needed to see, for example, that the standards of merit and of personal advancement cannot both logically be used at once to determine whether I or another man who is more highly qualified should get a job for which we are competing. Often these inconsistencies are first made evident when it becomes painfully clear in the course of conduct that we are paying too great a price for something we want—that is, when we discover on reflection that the money we spend could better be used for other commodities, or that the energy and time we expend could realize other values, or that the means we employ destroy other values.

Arguments over values, means, and ends are often confused as to what precisely the issue is: disagreement about means may masquerade as disagreement about ends or values, and vice versa. Consider two men who agree about the facts of a career open to both of them. It is a career which offers a good income, security, steady advancement, and a sedentary life without physical danger. If the first man values peace and security and the second man values excitement and novelty, then the first may regard the career as good and the second as bad. The facts are clear, but no single and exclusive judgment of value can be derived from them because the attitudes and preferences of the evaluator are a part of the ground of his decision. Now suppose *A*, the man who regarded the career as good, understands the character, personality, and social circumstances of *B*, who rejected the career. It is perfectly consistent with his own choice of that career for *A* to decide that it would be bad for *B*. And vice versa.

A third man, concerned with study of the other two, could, if he had enough information, predict with considerable accuracy the decisions each of them would make. But neither he nor they could assert with any reasonable justification that a life of security or a life of adventure was better *per se* than the other, and that one choice rather than the other was therefore better. Such a conclusion could be offered with justification only on the basis of an agreement that some further value, to which one choice but not the other led, was in itself good. So if all three valued longevity they might agree that one career was to be preferred, but if they valued intensity they might choose the other.

If there were no agreement on a further value, the three might decide that a life of security is better for men of one kind and worse for men of another. Analogically, when two cultures are confronted with the same alternatives, it is entirely reasonable to urge different choices on each of them. For the nature of the cultures and the material circumstances that surround them are relevant to the choice they ought to make. Indeed, one is unreasonable if he urges a choice without taking such matters into consideration.

Whenever a dispute turns out to be about fact, agreement can be reached—whether or not it can be reached about values—because facts are ascertainable, at least in principle. Suppose two men agree that the following things are valuable and perhaps even assign the same relative weight to each of them: a high standard of living, full employment, and a distribution of income without too great a disparity between the highest and lowest. But they disagree about the social and economic behavior which would bring these things into existence. One might advocate capitalism and the other socialism. The people involved in such disagreements may think they disagree about values. But the disagreement is actually about facts, because, if it can be established that one economic system rather than another yields the valued situation (and conceivably that can be established), then both men, if they are reasonable, will agree.

Now let us pose two examples. In the first, disagreement about values counts more heavily in the decision than agreement about facts. In the second, disagreement about facts counts more heavily than agreement about values.

First, suppose that someone accepts as fact that the mentally subnormal can contribute nothing to society and that, if kept alive, they are a drain on public finances. He yet may disagree violently with the conclusion that all feeble-minded people should be killed. He may prefer to keep them alive at any amount of public expense.

Second, suppose a man believes that "Aryans" are more valuable than people of any other "race," and that the more valuable should be treated better than other people. Then he learns a number of matters of fact: that innate intelligence is not correlated to ethnic origin; that Jesus, Shakespeare, and Leonardo were not, as claimed by many, "Aryan"; that the ancient Greeks were not the direct ethnic ancestors of contemporary Germans; and so on. He may conclude that he was wrong in his original high evaluation of the "Aryan."

It is clear that facts are relevant to values and may, in specific instances, be the most important factor in the choice of values. It follows that the methods and conclusions of science are important in evaluation. But intelligence in thinking of means and ends functions in two ways. Shrewdness, which perhaps brings "success," may consist in calculating the best means to attain ends, and possibly in being imaginative and ingenious about means. Wisdom may consist in appraising or evaluating ends and values.

Conduct of any complexity requires deliberate action, and that in turn requires reflection about means. We think first of the end we want, what John Dewey called an "end-in-view," and then try to find means to attain it. When we do, we must ask whether we are in a position to institute those means; if we are not, we think about those means as a temporary end, and seek further means that will allow us to attain them. So we reason from the end-in-view to its means, and then to means that will bring about those ultimate means, until we reach some means that we can institute now.

If a boy on graduating from high school decides that he wants to be a lawyer (end-in-view), he learns that to become one he must pass the bar examination (means). But he is surely not prepared to take the examination successfully now. What will prepare him for the examination (temporary end)? Attendance at a law school (further means). But he cannot enter a law school (more immediate temporary end) because most law schools demand college training for entrance (more immediate further means). So he must go to college, which he is ready to do now.[4]

So far we have only a hypothesis. We are saying, "If we do so and so (the means), then the end will come about." Our hypothesis that a set of means will attain an end is, like all hypotheses, a prediction which may be true or false, and that prediction is tested when we try to carry out our belief in practice. If our means do not yield our end, our prediction was false; if they do, our prediction was probably true. We can, as in the case of any empirical hypothesis, assert no more than probability, because it is possible that some other conditions, of which we may be unaware, brought about the end. We may, for example, attribute a business promotion to our own merit and

[4] When we *act* on the basis of our calculations, we reverse the process of thought; instead of starting with the end-in-view and terminating with the action we can perform now, we start with that action and terminate our behavior with attainment of the end itself. The boy goes to college, then to law school, then takes the bar examination. If he passes, he becomes a lawyer.

industry, when in fact they went unnoticed and the real means was the plea of the boss's daughter.

So far, our description of the rationale of means and ends is an analogue of scientific method, or an application of that method to the problem of attaining our ends. To apply it well is to be shrewd and perhaps successful. But does the analogue, or the application, go further? Can we also use it to appraise our ends? If we can, we will not only be able to add wisdom to shrewdness, but we will have more chance of lasting success. Most people bitterly oppose the maxim "The end justifies the means," for they take it to be an excuse for immorality. So they find it no justification of murder to say it was a means to a good end, like comfort, or advancement, or peace of mind. Yet good ends do justify means. What else could?

The feeling that the maxim is wrong is based, nonetheless, on a correct intuition. Ends justify the means if, and only if, *all* the ends those means yield are good. *The* end we aim at is only one of a number of consequences brought about by any set of means. Often the hypnotic effect of desire blinds us to the other consequences, but we must become aware of them if we are to decide whether to institute the means we think will yield our end. And we cannot conclude that the means are justified by our good end alone.

In appraising any end-in-view we must first ask whether there are, or readily can be, any means at all that will yield it. If there are no such means now, but may be later, we must put off any attempt to attain the end to a more suitable time. If there will probably never be such means, we must characterize the end as utopian and, if we still desire it, it must be as a counsel of perfection, a standard for judgment, or a fantasy. Second, we must ask how great a price we pay for the means—in energy, time, money, loss of friends or reputation, and so on. Further, is the end worth the price of the means? What other things could we do with that time and money, with those friends or that reputation? Is this end so much better than the others?

Then, what consequences will follow from this end once it is attained? Will they destroy the end itself in a short time, as Hitler's Reich was destroyed by the things it did? If not, what of their effect on other ends we desire, and on our values? Finally, what are the other consequences of the means we propose to employ? How consonant are they with the continuance of this end—granted we want it to continue —with other ends, and with our hierarchy of values? If we decide against these means are there others, more justifiable by their consequences, that we can use instead?

Of course, we do not have to ask all these questions of every deliberate action; there is often too little time, and usually only a few questions are relevant. But in principle all the questions can be asked, and in practice most of them are asked over a period of time, some of one proposed activity, some of others. At least if we are rational! And although there is rarely time for many of the questions, habit and mental sets about *types* of action can, as we shall see, prepare us to deal speedily with those types we have thought about in advance.

We can also appraise values and decide whether we ought to continue to hold them, that is, whether they are valuable. The ends that we desire may all be classified as instances of one or another value, although it is not always easy to know what that value is. We suggested that wanting a salary increase may be an instance of the value *wealth;* yet it may be an instance of the value *comfort* instead, if the man who wants a particular salary increase would reject a larger one because, perhaps, of the problems it would bring. A man who risks his life to save others may value sacrifice, or courage, or dangerous living, or reputation, or any of a host of other things. And it is not easy to tell, even for oneself, whether failure to live up to a professed value means that it is not really held or that it is too difficult to realize it, though one would like to.

When we are fairly clear that we do hold a given value and we derive from it a set of ends which are instances of it, we may regard the idea that these ends are desirable as a hypothesis to be tested. The tests take place in thought and practice, and they are the tests we have already suggested for the appraisal of ends. Again there is a scientific pattern. If these ends are desirable, then they will not destroy each other, their consequences will be desirable, and so forth. As we test more and more of the ends derived from a specified value and find the ends desirable, we increase the probability that the value is valuable. If the ends turn out to be valuable only in part or not at all, we may modify or reject the value.

What we establish by this procedure, however, is not a universal truth, like a general law of science, except with many qualifications. We cannot show that any value is true, or correct, for every person at all times. All we can show is that this standard which can guide action ought or ought not to be a value for ourselves and for others like us in relevant ways, including other values, personalities, and the circumstances of life.

In practice, not only can we guide our lives more satisfactorily by the use of the patterns of thought we have described, but we can clarify and sometimes resolve conflicts of value. As we pointed out, seeming disagreements about value can often be shown to be about means or particular ends. And the analysis and test of value enable us to accept values we had hitherto rejected and to reject others we had accepted. In addition, we can understand which differences in value ought to exist: when, for instance, men or groups differ significantly in personality, circumstances, and so on.

There are many people who will not like this statement because they believe that we should all hold the same values. They assume, of course, that what is valuable is the same always and for everybody, and that each value is absolute, unqualified by other values. But examination of their claims and their conduct—or our claims and our conduct, for we are all prone to overgeneralize about values—usually discloses that they apply their universal and absolute values so differently to different people in different situations that the values are emptied of any real meaning. We profess to value human life, for example, but we legally take the lives of those who take life illegally, and we permit killing in self-defense. Then we have other values which sometimes impinge on the value of human life. Some of us would rather die than be enslaved; some would rather die than lose honor or reputation, a loved one, or a homeland. We fight wars and take lives by the million because we value autonomy or freedom or power so much more than we do life simply in itself and under all conditions.

Perhaps what we value is what we *think*, in the broadest sense, advantageous. And what is valuable, what we ought to value, is *actually* advantageous, in terms of the tests we discussed. Then it would be no more important to have identical values than identical tastes. After all, it would destroy variety and richness, lessen individuality, force us into a single mold. It is important only that people hold in common the minimal number of values necessary for the functioning of societies and groups. And if they want freedom to develop their own values and to guide their lives in their own ways, then they must agree on a set of social values which will maintain procedures that allow individual differences. Meaningful differences in personalities and ways of life depend on similarities that create a social climate stable enough to sustain those differences. People who disagree about politics and want to continue to disagree must agree about political methods of resolving disagreement from moment to moment, so that the nation can be governed and people can lawfully oppose the government.

It may be that the specifically moral, the obligatory, is essentially this area of necessary agreement, which always limits conduct based on personal and unique values, and takes precedence over those values whenever there is conflict.[5] If that is so, moral values are the only ones that we need think of as universal, at least universal within a society or group. But we do not have to accept the conventional obligations imposed by any particular group as correct, even for that group. We can still criticize morals by asking several questions. Are these obligations necessary in order that there be any society at all? Are they necessary to hold together people with the preponderant values found in this group? Will they make it possible for the members to realize their personal values?

One cannot expect the structure of a group, which embodies its moral values, to realize for its members their personal values. That is up to them as persons. The structure need only provide the conditions necessary for individual attempts to realize personal values. The Declaration of Independence states that men are entitled to life, liberty, and the *pursuit* of happiness, not to happiness itself. It envisages a state whose laws provide the conditions for pursuing happiness, whose laws, indeed, are necessary conditions for happiness. But the sufficient conditions are to be provided by each man's intelligence and effort.

Policy is a course of action extending over some period of time and encompassing a number of particular choices or decisions. Businesses adopt policies like allocating more of the annual budget for advertising, thus affecting a host of particular decisions about expenditure. Colleges adopt policies like admitting to the student body only better students, thus altering particular decisions about the admission of future applicants.

[5] The study of morals and values is either descriptive, and so a part of social science, or evaluative, and so within the field of ethics, which includes both morals and values considered critically and prescriptively.

Policy stands midway between values and ends. It expresses a choice, based on our values, of a rule or plan of action which can guide particular choices, or decisions, *of a given type*. When we have a rule of action we can apply it to choices between concrete alternatives. Shall I vote for Burke or Hare, who seem equal in merit? Most of us have an implicit rule in such cases. When neither candidate seems better equipped for the job, vote for the one who represents our favored political party. Policy may apply to either means or ends. We may have a rule that, other things being equal, we will choose an end of one type rather than another. And although means are selected in terms of ends, we may follow a rule that leads us to decide for one type of means instead of another, especially when both types will bring about the end with equal efficiency and economy.

Although it is in large organizations that policy is most obviously indispensable to intelligent practice, it is also necessary in personal life. We are probably less explicit in the formulation of personal policy than a government is in its foreign policy. Nevertheless, we find ourselves again and again acting on the basis of a policy with respect to *types* of choice: we never give a beggar money, we are always polite to servants, and so on. Like any rule, policy may work individual injustice. One beggar may be deserving, for instance, while another is not. But if we did not formulate policies, explicitly or implicitly, we would have to stop to reflect about small matters of choice countless times a day, asking ourselves whether this decision or that accords with our values. We would then be incapable of accomplishing most of our ordinary tasks.

In formulating policies we are guided by our values and we choose types or groups of ends rather than specific ends. The businessman who values wealth may adopt the rule "Try to make a profit on every transaction"; or, if he is less shortsighted, "Be prepared to take a loss whenever it helps maximize profits in the long run." And these rules are tested, like values, by the particular decisions derived from them.

Social scientists are now consulted and employed by government in increasing numbers to advise on and implement its policy. Their training makes them specialists in knowing what types of means will yield given types of social ends. As scientists they are experts who may formulate policies to realize particular values or carry out policies made by others. But it is not their task to impose their own values. The values they must posit as government workers are the values expressed by the people at the polls and incorporated in government objectives.

This does not mean that as citizens social scientists should not urge their own values on the public, or that they should refrain from using their specialized information to appraise the values other people hold. One of the benefits of social science is that its findings can be used to test values. The social scientist in government can become one of the chief critics and teachers of the public. But he must never forget that the final decision about the values to be realized by government must be made by the people who are governed.

As an example of the relation of means, ends, policy, and values which involves most of the considerations just discussed, let us consider a man who wants political change. We will suppose he is a man dissatisfied with the *status quo* and desirous of realizing such values as justice, peace, and equality. The end he chooses, for the sake of his values, is a political and social order that does not at the moment exist. The means for attaining his end will depend in part on the existent political situation. If he lives in a monarchy, the end might be attained by influencing the monarch; if he lives in a democracy, by inducing the majority to vote as he wishes. But if there is no chance of persuading either the monarch or the electorate, he may contemplate seizure of power through violence. This will involve formation of a party or conspiracy, and policies to guide its particular choices.

There are a number of questions that a thoughtful revolutionist might ask himself before being committed to the use of violence as means. Will it be successful? If not, will the situation become much worse through the oppression, perhaps, of those who revolt? If successful, will those who oppose the revolt be so embittered that they will refuse to obey the new government, conspire against it, and, in general, be intractable? If so, will the seizure of power have to be followed by a reign of terror, or a suppression of dissident elements? How long would such despotism last? Could it be ended voluntarily when the opposition was eliminated, or would it breed a despotic bureaucracy which would perpetuate its own power? Are men who have been disciplined to a policy of violence fitted to rule? Would the other consequences of the means destroy the end?

The revolutionist would have great difficulties getting even tentative answers to these questions. He would have to look to history for information, and in his study of history he would have to be very careful that he took into consideration the special conditions under which past revolutions took place. The French Revolution, for example, was succeeded by a reign of terror and finally yielded to the Empire of Napoleon Bonaparte. The American Revolution might seem, in every way, more successful. But it would be important to note that the American colonists fought for independence from England, not domination over it, while the French revolutionists fought for the government of France.

And what about other values which may be involved? Does the revolutionist want to spend his own life almost exclusively in the service of the revolution and the new order? How many generations of people must he sacrifice before the new order can be attained? How can he decide that justice for unborn generations, if he can realize it, is more valuable than the happiness of people now alive?

The revolutionist may go ahead with his plans and learn by experience the consequences of violence in his own situation, or he may decide as a result of the questions he has asked and the answers he has found that it is better to give up his end than to pay the price for its attainment. But ends are not merely eliminated when it is decided that they are not worth their price. They are supplanted by others, and the elimination of this end plus the choice of another might lead to a change in the revolutionist's attitude toward his former value, justice. He might, for example, conclude that it is utopian and should be discarded; he might redefine it so that it could be realized without altering the face of society; or he might simply arrange it differently in his hierarchy of values, Where before it had been the highest of them all, now it might be placed below such other values as love and happiness. Yet whatever changes he makes in his values, he will find that his new evaluations are themselves subject to the same tests of reason and action as the old.

If one acts without the forethought displayed by our hypothetical revolutionist, he not only lessens his chance of success, but runs the risk of despair when he realizes what might have been. We all tend to accept without reflection the major values of our society, and we seldom try to find out how well they work for us personally. Even if we suspect that "a successful career," for example, as it is usually defined, will not accord with our personal values, we find that it takes courage and intelligence to resist the pressures of family and friends and the lures of acceptance and prestige.

There are many recorded instances of the anguish of "successful" men when they first manage to take stock of themselves and compare what they are with what, years before, they hoped they would become. They find that they have blunted their earlier sensitivity to experience, become callous in their response to the suffering of others, and lost the taste for ideas and the appetite for the arts. What they have in return, the material advantages of "success," the envy of "failures," and the companionship of their peers, seems vanity and ashes.

How can we avoid this sort of overpayment, this living up to standards which are not our own? It takes more even than intelligent forethought. It demands critical self-consciousness, a developed sensibility, and awareness of the commitments we make—for the most part implicitly—to ourselves and others. There is too often a step-by-step passage from one situation to another which leads us unwittingly to ends neither foreseen nor desired. So we need the time and courage for a periodic re-evalution of our ends in the light of our values, and of our values in the light of both means and ends. "There is a tide in the affairs of men which, taken at the flood, leads on to fortune." We must know the tide if and when it comes and be prepared for drastic change in our lives if it is warranted, for otherwise we know only when it is too late to act.

# 8

# Science and Free Will

Even if science cannot pronounce on the subject of moral conduct, it would seem to be relevant to a factual matter: Is man really free to choose? If science can answer that question, it can show whether moral conduct is possible at all. For unless men are free to choose between alternatives, they are not moral agents. Only if man is free to choose does he fulfill the necessary condition for the existence of morals. And if he is not free to choose, there can be no morals.

The scientific question is this: are all human acts completely determined? Whether the determinism is thought to be hereditary or environmental or a combination of the two does not affect the consequences for morals of our answer to the question. The distinguished psychologist B. F. Skinner believes that behavior is entirely determined, so he argues that "freedom" and "responsibility" are meaningless words, which we must learn to do without. After all, if we do only what we must, not what we will, we are not responsible for our actions in the same way we would be if we were free to choose them.

What are the consequences of determinism for morals, as some determinists see them? If all actions are determined by prior causes, then we are not entitled either to blame a man for what he does, or praise him for it. If he cannot be blamed, he should not be punished. If he cannot be praised, he should not be rewarded. As we examine our own conduct, it makes no sense to be proud of an accomplishment that was hard-won. We had to do it. And it is equally senseless to feel remorse or regret for some folly. It could not have been otherwise.

It is no wonder, then, that many people who accept the logic of the argument feel that the only way to oppose it is by denying the premise, determinism itself. For if these are the proper consequences of determinism, and if determinism is true, then man has been self-deluded for centuries, and the very terms in which he talks about himself are meaningless. Indeed differences between man and the machines he makes would be just that he is (as yet) more complex.

How can we best deny determinism in human conduct? Shall we appeal to our feelings of freedom? "Sir, we *know* our will is free," said Samuel Johnson, "and *there's* an end on't." But we "know" it only in the sense that we feel it. We feel that we may choose A or B and that we choose without inner compulsion. Is that evidence? Not for the determinists, who might point out that, when choosing, we concentrate on the alternatives before us. Suppose, instead, we looked back and asked why we chose as we did. We might find conditioning factors. And even if we couldn't find them, that wouldn't mean they weren't there.

Experimental psychology and psychoanalysis are of no help to the indeterminist, because they are concerned with finding determining or conditioning factors and admit no limits to what they may discover. So he has had recourse to physics. Here he rests his case on the Principle of Indeterminacy, stated first by Heisenberg in 1927, according to which we cannot determine precisely both the position and the momentum of a particle. The conditions of observation are such that the more precisely we determine either position or momentum, the less precisely we determine the other. Sir Arthur Eddington made much of this "spontaneous" behavior of the atom, and he has been followed by many who are less qualified to interpret science than he. "A complete determinism of the material universe," Eddington said, "cannot be divorced from determinism of the mind." And Heisenberg's Principle, he argued, shows that the material universe is not completely determined, because material particles behave unpredictably. Now, man's brain is made of just such particles. How, then, can all its behavior be determined?

What confuses this argument is the meaning of "indeterminacy." Determinism is the belief that all events are caused by prior conditions. Indeterminism denies that. Heisenberg's "Indeterminacy" does not assert that some events are uncaused, but only that we cannot measure them (or, to put it in language that better reveals two meanings of the same word, that we cannot *determine* their measurements). So we do not know that the atom behaves "spontaneously"; we only know that we cannot measure some of its behavior. The case against determinism is, so far, not proven.

The case for determinism has not been proven either. But it seems plausible to many people because of the great success of science in finding the causes of events, and because we all assume causation throughout daily life. Our actions are meaningful only because we expect means to yield ends. Still, the question that bothers us is not "Is nature determined"? but "Is man determined"? And much as some men hate to answer "Yes," others delight in it, and boast that they have given up wishful thinking and human illusions to face the stark truth.

Let us analyze the idea of determinism a little more as it applies to man, so we can see better what it means. There are four questions which should be kept separate because they may have separate answers. One is led to ask them by a standard determinist gambit. If an indeterminist says he is confident that he is free to do as he chooses, the determinist counters by asking if he is free to choose as he chooses.

So the questions are:

1) *Am I free intellectually to choose among the possibilities I conceive?*
2) *Having chosen (freely or not), do I permit myself to carry out the choice?*
3) *Do others permit me to carry out the choice?*
4) *Do I control the means to bring about what I have chosen?*

The fourth question is about the power at my disposal, for "control of means" is power. If I have too little power, I cannot make my choice effective. I may choose to attend the opera tonight, but I need the price of admission. If I don't have it, I can't attend. Perhaps "choose" is the wrong word in this case, and should be used only of real possibilities—i.e., where I have the power to do what I choose, and the question is what it is I choose to do. Then we might say of the case in which I don't have the price of a ticket that I would like to attend the opera tonight, not that I chose to do so.

The third question is one of social and legal freedom. In the ordinary sense of the word, I am "free" if I am not stopped from doing as I please. Here it is irrelevant whether or not my choices are determined. So long as they are *my* choices, I want to carry them out. Determinism has nothing to do with politics, or with human freedom at the level of politics. In that arena I mold my destiny in cooperation and competition with my fellows. The determinist thesis—as in the answer to the indeterminist above—grants that often I do just what I want to, and then asks whether I could have wanted to do otherwise.

The second question raises the possibility that I want to do something no one else stops me from doing, but that I cannot make myself do it. Equally, I may not want to do something that no one forces me to do, but I am internally compelled to do it anyway. Are there really such possibilities? Apparently there are. But am I entitled to say I want to do something or have chosen to do it if I don't in fact do it when given the opportunity? Yes (although there may be some ambivalence in the desire), when I suffer from a psychic disorder that prevents me from doing some things I want or makes me do other things I don't want. If I become tongue-tied when I want to propose marriage, or feel impelled to steal a trinket I don't like, and suffer an agony of shame over it, I am not acting in accordance with my intellectual choice.

The virtue of this example is that it points up a particular kind of lack of freedom. The tongue-tied man and the kleptomaniac are not free to do as they desire, not free to carry out intellectual choice, although no one else interferes with them. And how does this bear on determinism? The determinist finds us all equally unfree because we are (he thinks) equally conditioned to do what we do. His thesis does not permit him to make a distinction where there is a clear difference, as there is here. A man who decides to steal and becomes a thief is free to do as he chooses. A kleptomaniac who decides not to steal any more but does so anyhow because of internal compulsion is not free to do as he chooses. Even on the determinist thesis, then, some men are freer than others.

The first question should be at the heart of the determinist argument. How free am I in terms of sheerly intellectual choice? There is a common opinion that no one else can interfere with what we think, only with what we do. But the opinion is false. We can be brainwashed, persuaded, lied to, and deprived of information. Clearly, on this level, too, some men are freer than others. If I sit down quietly in my own home to reflect about a matter, I am freer in my judgment than if I am in a prison cell, confronted by captors expert in "thought control."

At this point, we can probably conclude that the thief, who steals because he wants to, is morally responsible for his thefts, while the kleptomaniac is not morally responsible. We can probably conclude, also, that the man who makes up his mind in peace is free in a sense that does not apply to the victim of brainwashing. After all, it is one thing to make up my own mind, no matter how thoroughly conditioned my thinking is, and quite another thing to have *you* make up my mind —by drugs, hypnosis, persuasion, or whatever.

But there is still more to freedom of intellectual choice. The determinist denies there can be such freedom because, he says, every act of thought is caused by something that went before it, making the mind think this way rather than another way. The indeterminist means to deny such causation, but he cannot, except in sheer folly, deny that there are reasons why we think in just the way we do. "Cause" and "reason" are very different words here. Cause is in the past, reason (in a way) in the future. At the fork in the road, I may know no *reason* to go left or right. I choose to go left. Why? The *cause* may be left-handedness, or habit. If, however, I knew that the road to the right was a swifter and smoother way to where I was going, I would have a *reason* to go right. For the determinist, there are causes why one man accepts some kind of reasons and why another man does not. But suppose a man, for whatever causes, chooses to abide by the principles of logic and the rules of evidence. Isn't his thinking dominated more and more by reasons (the facts and arguments before him) and less and less by causes (which are in his own past)? Doesn't the attainment of rationality, therefore, mean some escape from the bondage of irrational conditioning? There is some freedom, surely, in accepting conclusions as true because they seem warranted by the evidence, rather than having to accept as true only the conclusions you wish were true.

Perhaps another context will clarify and extend the argument. There is a mistaken identification of determinism with fatalism. Determinism is the belief that what does happen must happen because all events are effects of preceding causes; all actions, thoughts, and emotions are the result of heredity and environment, which make each man just what he is. Fatalism, the belief that what will be must be because all ends are fixed in advance, *seems*, on cursory analysis, to be little different in meaning.

So some people have contended that determinism leads to fatalism, that the only difference is one of emphasis. Determinism, they say, looks at the causal chain in a temporal sequence from cause to effect. Fatalism starts by proclaiming the necessity of the same ultimate effect and secondarily looks back to the cause that brought it about. Then, the argument runs, since the strict determinist would say that any event that *in fact* occurred, *necessarily* occurred, he is confronted with the question why our empirical knowledge is only probable, as scientists claim, rather than certain, as it should be if events do occur necessarily. He answers that probability is only a sign of the limited state of our knowledge today: if and when we know enough, empirical knowledge will be as certain as analytical knowledge.[1] Thus the determinist rules out the possibility that empirical knowledge can *never* be more than probable because chance exists in some measure in the universe. It is this denial of the possibility of chance that is regarded as equating determinism with fatalism.

The argument overlooks a real difference in the two beliefs. Fatalism insists that an event *must* occur, *no matter what* preceded it. If *E* is fated to occur, then even if it has been preceded by conditions which ordinarily cause non-*E, E will* occur and non-*E* will not occur. So fatalism is antiscientific. Determinism, on the contrary, accepts the inevitability of the event *only if* it was preceded by other events which caused it. Determinism is a possible interpretation of the world described by science and, right or wrong, it tries to base itself on science. Determinists say: *if* and *only* if the cause exists, does the effect exist.

If fatalism and determinism are different in theory, they should be different in practice; and an attempt to apply them does, in fact, show that they are. If a man passes the bar examination, he will be licensed as a lawyer. He will not be licensed, unless he has passed other examinations as well, as a physician, engineer, or aviator. But according to fatalists, if that man is fated to be an aviator he will become one, examined and licensed or not. Since in case after case such predictions are refuted, fatalism is proved false. And if it is still defended, that is only because the fatalist may admit no conceivable con-

[1] Probability is not, according to the best scientific thought, a measure of human ignorance. It is a measure of the frequency with which events of a specified type occur. There is a probability of one-sixth that the number three will be face upwards when we throw a single die. That means that three will show with a frequency of one in six times if a die is thrown endlessly. It does not mean that we can calculate probability for a single, or unique, occurrence.

ditions under which it could be tested, thus violating the verifiability criterion of meaning. For if the man we have been discussing lives his life as a lawyer and dies without ever becoming an aviator, the fatalist says that *he* was wrong in thinking this man was fated to be an aviator. Therefore there is no test of the basic fatalist hypothesis, that our ends are fated, but only of whether or not we know what is fated.

If determinism is a scientifically meaningful hypothesis it should, in principle, be capable of test. If we were to discover an event which had no cause,[2] we would say that the determinist hypothesis was refuted. Thus, while fatalism is either obviously false or else tautological, determinism is empirical. Of course, if an uncaused event should occur, some determinist scientists might still hope for ways of finding entirely new types of causation, which they probably could not even describe. But until they found them, determinism would be for them an article of faith, as fatalism is for others.

Now, does a totally determined world eliminate freedom and many other things as well, such as chance or accident? It does not.

Let us consider chance first. If all things are determined, it is nevertheless the case that only omniscience could know everything that has happened and so predict accurately all future events. Suppose I were to discard my only toothbrush, which I had used too long, and had forgotten to buy another by bedtime. Suppose also that my conditioning makes it impossible for me to go to bed without brushing my teeth and, in consequence, I dress and go to the nearest drugstore for a toothbrush. Now let us assume a Mr. Hyde whose wife has left him. He is in despair and, due to his conditioning, is led to drink and careless driving. As I cross the street to the drugstore, Mr. Hyde passes a red light in his car and kills me. All was determined, but only from the standpoint of omniscience, for from the perspective of my chain of cause and effect alone, and of Mr. Hyde's alone, the incident was chance or accident. From knowledge of my life alone one could predict only that I would dress and go out to buy a toothbrush; and from knowledge of Hyde's life alone only that he would drink and drive carelessly. One would have to know both our lives to predict both our actions, and there would be no reason to know them both, or to

[2] Not an event whose cause is as yet unknown, for we might later discover that cause, but an event that occurs contrary to predictions that were based on ordinarily sufficient evidence.

know them rather than any other two lives, because there is no reason to believe they would ever affect each other.

Determinism is often erroneously thought to be opposed to human freedom. In a totally determined world, we are told, there is no real choice; we may *feel* that we choose freely, but that feeling is an illusion masking the necessity for choosing as we do. If this were true, would it follow that in a world partly determined and partly undetermined, the more the determinism the less the freedom; and that in a world totally undetermined, freedom would be total? It is often assumed that these consequences follow, but it can probably be shown that they do not.

The particular problem of choice when we are confronted by alternatives seems to be: Which do we prefer? But important matters are rarely so simple. The alternatives are not likely to be things which we want in themselves, like a dessert or a cup of coffee, but directions, or means which lead to ends we desire, like pursuing a course of study in a school of law or a different course of study in a school of medicine. In a totally undetermined world it would make no difference what we chose as means, for there would be no reason to suppose that any means would function as a cause to yield an effect. In other words, no action could be regarded as a means because it would have no consequences. So there is no point in choosing at all if it is entirely a matter of chance whether or not we get what we aim at; we might as well make no choices and take what comes. Choice, we may conclude, is only meaningful in a world which is at least partly determined.

There is a common belief that many actions of determinists are inconsistent with their doctrine. It is argued that if a determinist criticizes someone's action, he thereby gives the lie to his own belief that that action was determined, and that the man in question could not have acted otherwise. If a determinist wants to jail a criminal, is he not implicitly assuming what determinism denies, that the criminal is responsible for his crime, in the sense that he might have chosen not to commit it? Not at all. One answer a determinist might make to these charges is that other men do the things they are conditioned to do, *and so does he*. In these cases he is so conditioned that he criticizes one man and wants to jail the other. He could then point out that the examples urged against him presupposed that he believed all other people to be determined but thought himself undetermined, which, of course, is not the case.

But there is another explanation of the behavior of the determinist who criticizes an action or wants to jail a criminal. If the determinist says that a man cannot control his actions but does what he must, he does not mean what a fatalist would mean, that the man must perform this action no matter what has gone before. The determinist means that the action is performed *because* of the way the man has been conditioned, and that had there been different conditioning, a different action would have resulted. And there can be reconditioning.

The determinist himself, as we have seen, may be so conditioned that he criticizes another. If we ask what the use of that criticism is, he may answer that it is a new element in that other man's environment which may itself condition him to behave otherwise. In fact, it is only if the man can be conditioned by what happens to him that criticism can be effective. In a completely undetermined world, however, criticism could not be expected to alter behavior; it would be only a gratuitous expression of attitude, made for its own sake. Jailing a man would be sheer revenge or punishment or sadism, without the excuse that it might change the prisoner's behavior in the future.

Finding a place in a determined world for accident and for significant conduct with respect to others is still not finding a place for freedom. But the argument so far is the foundation for an interpretation of freedom. Conditioning is never complete; there is always further conditioning and reconditioning, the destruction of conditioned patterns of behavior and the acquisition of new patterns. A child is conditioned to accept the moral values and attitudes of the adults about him. Then he goes to school and much of his past conditioning is reinforced. But at some point in his education he may learn to think logically, to demand evidence, to question. This, too, may be called conditioning. The new conditioning destroys much of the old, and understanding the nature of conditioning permits deliberate change in oneself and others. A man who is dissatisfied with himself, perhaps because of new things he has learned, may decide to alter himself in specified ways. How can this be done? By discovering what changes in the environment would bring about changes in himself, and by instituting those environmental changes. More generally, the conditioning of human behavior is made up of a great variety of external forces, some of which run counter to others. Learning the ways in which we are conditioned permits a deliberate manipulation of those forces to yield desired ends, to alter the behavior of oneself and of others. This is surely as persuasive a description of freedom as any we can find.

There is still one rather obvious objection to what has just been said. Even if a man can use the causal patterns of his environment for his own purposes, is it not true that those purposes are themselves the necessary results of causation? And does that not make freedom illusory? The answer is no. Choices can be made blindly as a result of past conditioning of which we are unaware; or they can be made in knowledge of our own conditioning and with reasons for choosing one alternative rather than another. Knowledge of our own tendencies may be thought of as itself the result of some previous conditions, but it helps free us from other conditions which would lead us to thoughtless choice. Awareness of the consequences of alternatives clarifies the meaning of each choice and permits us to choose for reasons, not merely because of our conditioned tendencies. The distinction between the two kinds of choice is a vital one and should not be blurred because the word "choice" is used for both.

A simple analogy on the mistaken use of another term, "selfish," may make our distinction between two kinds of "choice" more persuasive. Hedonists have insisted that all human action is motivated by a propensity to seek pleasure and avoid pain. No matter how different the things in which Cotton and Mather take pleasure, their behavior is fundamentally the result of the same drives. Cotton may want wealth and Mather prestige, but their behavior is equally "selfish" in the sense that it seeks to gratify the self. If, instead of the imaginary Cotton and Mather, we think of the Roman emperor Nero and St. Francis of Assisi and we call them both selfish since they both did what they wanted, we will need still other words to distinguish the kinds of things Nero wanted to do from those Francis wanted to do. Then we might say that Nero's was an egoistic or selfish selfishness, whereas Francis' was an altruistic or unselfish selfishness. What is the advantage in retaining the word "selfishness" for both men's conduct when the difference is so great that we must use contradictory adjectives to describe each type? Why not just characterize them as "selfish" and "unselfish"?

There is so great a difference in the ways in which we choose, and consequently in our choices, that it is valuable to use one word for unreflective choices and another for reflective choices made in the light of knowledge. The former are bound and the latter are free.

# 9

# Communication, Symbols, and Society

Communication is a subject larger than precise meaning and literal exposition. It includes other uses of language, and nonlinguistic communication—gesture, ritual, symbol, action, art, and religion—in so far as they have communicable meaning. In dealing with this aspect of communication, its relations to man and society are all-important. Man can be defined as a symbol-using animal at least as well as he can be defined in any other way; and social relationships can be traced effectively on the basis of the network of symbols and communications in society.

Communication is more than meaning, for it is *meaning* that is communicated. Meaning may be private, imparted to no one; but communication implies at least two minds. An identical meaning may be communicated in several different ways: by speech, writing, or gesture, for example. *Communication is a process in which meaning is conveyed. It is also purposive: it is an intended provoking of response.* If I see dark clouds in the sky I may expect rain, and it is this I refer to when I say that I know the meaning of dark clouds. When I think of dark clouds as meaning something, I am treating them as a "sign" I am interpreting. But I have not received a communication because no one intended to provoke a response by placing the clouds before my eyes.

Communication is always social. It contains three elements, one cf which has an important corollary: (1) a person who comunicates, (2) the signs by which he communicates, and (3) a person who interprets the signs. The communicator need not be present, or even alive. Dead authors and artists communicate. The signs must be present to the interpreting mind, although they need not be present in fact. They may be only remembered, as the first chapter of a book is when one is in the middle, or the first movement as a symphony nears its close.

The corollary is implied by the third element. To be able to interpret a sign one must know the conventional system within which the signs have their meaning. One must know English in order to understand an English sentence. To an Eskimo visiting one of our cities for the first time, a green light at an intersection may have no meaning, or it may not have the meaning intended by the highway commission; one must understand the system of traffic signals in order to interpret a green light successfully. In the same way, one must be familiar with the principles of Chinese music in order to grasp the meaning of a Chinese musical composition; one must understand the system of Japanese flower arrangements in order to know what is intended by the flowers set in a Japanese room.

With this corollary, what seemed simple and straightforward becomes complex and devious. The conventional meaning system may be one that gives systematic rules for interpreting signs. Traffic signals and mathematics do so, mathematics less successfully because there are cases of context altering meaning even in mathematics. But a language, especially a language spoken by a civilized nation, is subject to all the difficulties we discussed under the heading of scientific language, and many more. Both language and art have implicit in them the presuppositions of the culture from which they come. Their use reveals much of the intellectual history of that culture and the psychology of the people who compose it. Language and art, in turn, help make that history and that psychology what they are. We will try to explain these perhaps cryptic statements.

*Community* and *communication* are words which show an immediate similarity. They emphasize commonness, togetherness. People gather or live *together* for certain purposes, and they *share* meanings and attitudes; the first presupposes the second, for without communication there is no community. Community depends on shared experience and emotion, and communication enters into and clarifies the sharing. Forms of communication like art and religion and language are themselves shared by a community,[1] and each of them contains and conveys ideas, attitudes, perspectives, and evaluations which are deeply rooted in the history of the community.

Language shows the differences in cultural ideas most clearly. Malinowski insisted that no dictionary translation into English of the speech of Trobriand Islanders would give their meaning, because their language carried traditional ideas unfamiliar to people outside their culture. So a seemingly simple description of arrival in a canoe carried in its very syntax and its metaphors a great weight of competitive boasting. "We paddle in place; we turn, we see our companion," may sound like description, but for the Trobrianders the emphasis is on being in front, being able to paddle faster than others.

The syntax of modern English expresses many deep-rooted ideas of Western civilization and makes it difficult to say untraditional things. The way in which English expresses the idea of the self in terms of ownership is a case in point. I *have* a body; I *have* a mind; I *have* a soul. Then who am I who has all these things? I may be thought to be one or another of them, or all three in combination. The syntax, however, makes it sound as if I were none of them, but something else, and what else could I be? This confusion is due to a traditional metaphysic of substance and attributes, according to which things referred to by nouns or substantives are somehow substantial, while their attributes, or properties, or qualities, are not. Of course, "body," "mind," and "soul" are ordinarily substantives but they are used as if they were attributes, or properties, of "I" in the instance above.

As we learn our language, we acquire the cultural ideas with which it is freighted without conscious realization that we are doing so. To take a different kind of example: God, obviously not endowed by Christianity with sexual qualities, is always "He." His Son is born of a woman and God is referred to as the Father. The images of a patriarchal society thus color our religious attitudes through the very language which conveys them.

Language and civilization grow together. As other aspects of the civilization develop, its language, whose structure already expresses an earlier age, must be wrestled with and sometimes altered in order to express new thought properly. But new thought does not grow apart

[1] Art and religion, as we shall contend, are not only forms of communication; they have aspects of creation and feeling which are individual and can be thought of in themselves.

from language; new thinking and new ways of speaking go together. One way of expressing new ideas, especially abstract ones, without wrenching the language from its customary uses, is to use metaphors based on more familiar notions, thus assimilating the new ideas into the older linguistic and intellectual habits.[2] Some ideas of physics, for example, were so difficult for even physicists of past centuries to understand, that they immediately acquired metaphoric statement. How would one say in literal English what is conveyed by "bodies attract and repel each other," which sounds for all the world like human affection and aversion?

In contrast there is the difficulty faced by a thoroughly literal man with scientific training when he tries to understand the humbler uses of language. These have their meaning defined by the human behavior of which they are part, not just by the words that are uttered. "Blood is thicker than water" is not intended as a biological truth. Let us first distinguish three different situations, all of which are meaningful, so that we can isolate the "humbler" expressions and see their value. (1) A man carrying a pail of water under whose weight he is stooped and straining shows the observer that the pail is heavy. (2) If the same man utters the statement "This is heavy," we have that information without watching him. (3) If he is a scientist, he may make an exact statement, "This pail of water weighs fifty-seven and a half pounds." There is an enormous gain in the scientific quality of communication.

Is this gain without loss? No; much has been lost. The more exact statement, which specifies the weight, has lost the idea that it is a hardship to carry this pail of water, an idea that was conveyed by the first statement. And even the first statement has lost the *quality* of hardship that we feel when we see this man carrying this pail over this ground. Battlefield communiqués are more "scientific" than firsthand descriptions by soldiers and they convey specific information about a battle more effectively, but there is a loss as well as a gain in this advance in accuracy. For when the *experience* of battle is *reproduced* in literature, as it is in *War and Peace, The Charterhouse of Parma,* and *The Red Badge of Courage,* more is kept of the meanings and power of experience than can be found in statistical tables of Napoleon's campaigns or of the American Civil War.

[2] See pp. 22-25.

The statement "This is heavy" is an example of a "humble" use of language. It is imprecise—for we don't know how heavy it is—and perhaps querulous. The hardship of carrying the pail of water is not stated; if we did not know what the situation was we would not know that the man who carried the pail was complaining. If he had been discussing a brick of solid gold the statement "This is heavy" might imply happiness if he were the owner, or envy if he were not.

Many "humble" uses of language may be thought of as non-cognitive, in the sense that the words alone do not convey the meaning of their use. That meaning is in the situation as a whole and the words, though central to the situation, cannot be disconnected from it and still retain their function. Here we are thinking of the use of words in attention-getting, reassuring, sharing, dedicating, and things of that sort. We often use a person's name at the beginning of a sentence addressed to him, to make sure that he listens to what we say. Sometimes we utter whole sentences merely to make him aware of our presence and so to assure ourselves of our importance, perhaps even of our existence. "What a nice day," we may say, just to get attention, to be recognized. At other times we speak in order to give another person assurance that we are aware of his presence, saying perhaps the same thing that called his attention to us. Again, we reassure him by saying any of a variety of things which shed no light on the matter which has led us to speak. We say, "Everything will be all right," or "There, there," or "Don't worry about a thing." By expressions like "My, my," or "Did you see that?" or "Isn't it wonderful?" we show that we are sharing an experience and often increase the pleasure in another person's experience, and in our own, by emphasizing the fact that we are undergoing it at the same time.

What we have been saying about language should show its wonders as well as the difficulties of using it for thought. But the wonders far outweigh the difficulties. In fact, the difficulties are only the problems of mastering language for the service of thought. Without language, man is scarcely human and thought only rudimentary.

Man is not the only creature capable of co-operation with his fellows for the sake of shared ends, but he is the only creature who has culture. Some "higher" animals co-operate in a rudimentary way and, under laboratory conditions, have been trained to co-operate even more. It is in insect societies, however, that co-operation is at its height in the nonhuman world. The ant, the termite, and the bee are endowed with instinct which results in highly social behavior. The ant and the bee communicate sufficiently so that one member of the heap or hive can direct the others to food. The dance of the bee seems even to communicate an estimate of the direction and distance of the food that has been found. Nonhuman animals, other than insects, do not seem able to communicate with such exactness; but they are able, by vocalizations of different kinds, to express emotion and to attract the attention of their fellows. This often results in behavior directed to their needs or wants.

The ant communicates by touch and the bee by movement in space. Animals gesture as well as make sounds. One has only to watch a chimpanzee in the zoo to see a stance of hostility or a movement of conciliation (terms for human behavior, to be sure, but metaphorically descriptive enough). Yet communication among "higher" animals is scarcely greater than among the social insects—if, indeed, it is as great. And although there are herds and packs of animals, there is nothing in the animal world below the level of man that remotely resembles insect societies in complexity of organization. The reason seems to be that societies may exist on the basis of instinct or of language. Insects have the former and men the latter, and other animals do not have enough of either.

More instinct-directed behavior is found in creatures who are lower in the evolutionary scale. A rough rule is that the higher the intelligence of the species, the less its instinct. There are wasps that fight the much larger tarantula spider, sting him in a spot the size of a pin point, which paralyzes him, and carry him off to a nest where they deposit their eggs, leaving the living flesh for the larvae to feed on. And a wasp separated from her fellows at birth and confronted with a tarantula for the first time behaves in the same way. The sea exhibits countless instances of marvelously developed instinct. Grunions lay their eggs in the sand just as the tide is receding. The adults get back to sea, leaving the eggs to mature in the sun-warmed sand until the returning tides rupture the eggs and carry out to sea the baby fish which emerge from them. European eels swim some three thousand miles from their fresh-water homes in rivers and streams to the Sargasso Sea, where they reproduce and die, leaving their young to return to fresh water, a trip taking as much as three years, where they live until sexual maturity and then repeat the cycle.

Either man has no instincts at all, or we do not know what they are. His chief internal equipment for survival is the brain and central nervous system; his chief external equipment is his hands. Brain and hands create tools which are a much extended and variable external equipment. Other animals are limited to the equipment they carry with them as parts of their bodies: fangs and claws, tusks, paws for digging, quills or shells for protection. These alter slowly in the course of evolution but they cannot, like tools, be discarded or created at will.

In addition to brain and hands, man's equipment for survival includes those organs, like the larynx, which make it possible for man to live socially. Man has not survived, as the lion has, individually, in isolated families or small herds. Man is able to talk, and speech makes possible co-operative undertakings which protect him from his enemies, shelter him from the elements, and lead to the creation, the transmission, and the alteration of culture.

Only man has a developed, or verbal, language, so it is only man who can learn things precisely and store them in memory in all their essentials. Language is a necessary condition for the existence of culture, for what is learned by one man is transmitted to others; and what is wanted for co-operative undertakings of any complexity can be expressed in words. Then the organization of a society and the things it has learned can be transmitted to the next generation, thus continuing the culture that has been created. Language manages what nature cannot: the transmission of acquired characteristics. Each generation adds to the store of culture, or changes what it has inherited, making culture cumulative. Growth and change give man a history, and, properly speaking, he is the only creature who has a history.

Finally, language contributes inestimably to man's sheer physical survival, not only by giving him knowledge acquired by others at his own time or in the past, but also because language is needed for co-operation to meet novel situations. Instinct alone allows the insect to cope excellently with a limited number of situations that are invariant, or nearly so; language may be used to guide behavior in new ways in the face of the unexpected.

There are other forms of communication which, as we shall see, store and transmit knowledge, hold a society together, and direct conduct. But it is doubtful whether these would even exist if man had no language. And surely only language can communicate precisely the abstract ideas of science and philosophy or the practical wants and concerns fulfilled in industry, trade, and the countless minutiae of life.

In the life of the child, language often has a part contrary to the one it plays in the development of civilization. In civilization, verbalization and conceptualization often take place *after* the event. Religious ritual is explained by the theology that follows it, political action is studied and its principles *subsequently* formulated, Aristotle's rules for the drama are written when the great age of Greek tragedy is past. The child, however, is often taught the word *before* he knows its referent, and may even be able to repeat sentences whose meanings emerge only in later life.

We cannot specify exactly the part that language plays in socialization, but we can be confident that it is a very large part. Socialization not only makes a person a member of the culture, but it also makes him fully a human being. Apart from society, man does not even exhibit potentiality for living in the most rudimentary culture. We cannot be sure precisely how much the humanness of man depends on language, but there is a good deal of evidence that without it man is scarcely recognizable as human, and that language is probably necessary for all higher cultural accomplishments.

There have been many reports of the survival of children who were lost at a tender age in uninhabited places. Only a few of the reports are credible, but these describe their subjects in very much the same way. Strange man-like creatures found in isolated regions in the Middle Ages, children presumably raised by wolves, and the few feral men reported more recently are regularly described as savage, unintelligent by human standards, and exhibiting more characteristics of other animals than of man. In some instances the children never can learn to speak or write or to take their places as ordinary members of society. When found, they usually go about on all fours, are savage to men, and prefer raw meat. Occasionally one of them lives for some years in captivity and acquires a few of the habits of the culture, including even a simple vocabulary. There is great theoretical importance in reports of feral children (and in the case cited below) because the decisiveness of the role of language in socializing children is one of the hypotheses on which social scientists can scarcely experiment.

In the past it was believed that behavior is learned to a great extent by imitation. But the occasional case of the child raised in the company of one or more adults who do not speak indicates that the special factor of language may be more significant for socialization than had been assumed. This is evident in the story of Isabella.[3] In brief, the sociologist Kingsley Davis studied a girl called Isabella who had lived the first six and a half years of her life in a single room with her deaf-mute mother. She rarely saw the rest of the family and in consequence had learned no language; she and her mother communicated solely by means of gestures. When Isabella was found and brought into society she showed some of the characteristics of the feral child; according to Professor Davis, "Her behavior towards strangers, especially men, was almost that of a wild animal, manifesting much fear and hostility. In lieu of speech she made only a strange croaking sound. In many ways she acted like an infant."

Although there seemed no hope for Isabella, she was trained intensively, at first with almost no success. It took one week to get Isabella even to attempt vocalization. Yet little more than two months later she was putting sentences together, and sixteen months after that she had a vocabulary of nearly two thousand words and was, of course, reading and writing. Her I.Q. had trebled in a year and a half and she was well on the way to catching up with children of her own chronological age.

Since the chief lack in Isabella's socialization up to six and a half years of age was language, one should be entitled to conclude tentatively that it is the absence of language in the environments of feral children that accounts for much of their seeming nonhumanity. In other words, language is a *necessary* condition for *rudimentary* socialization. Yet language alone does not bring a child up to the average level of his society; it is not a *sufficient* condition for *average* socialization. This is evidenced by studies of children in remote, relatively unchanging areas, children who were raised normally on the spoken word and were given ordinary education. One such study, of isolated mountain children,[4] shows that their average I.Q. is less than the average of children in ordinary small towns. It is important, however, to note that whereas the scores of the mountain children on intelligence tests were consistently a little below average for the nation, Isabella's score was almost zero. This is a measure, perhaps, of the part language plays in socialization.

[3] Kingsley Davis, "Final Note on a Case of Extreme Isolation," *American Journal of Sociology,* Vol. 52, 1947, p. 432.

[4] Mandel Sherman and Cora B. Key, "The Intelligence of Isolated Mountain Children," *Child Development,* Vol. 3, 1932.

In its simplest definition, a symbol is a conventional sign. A sign represents, or signifies; it conveys a meaning. Signs may be either natural or conventional. The meaning of natural signs is *discovered,* for they are part of the causal connections of nature. When litmus turns red in a solution, the color is a sign of the presence of acid. The meaning of conventional signs or symbols, however, is fixed by use and agreement; they are arbitrarily chosen, in that they have no causal connection with what they represent. All words are symbols although not all symbols are words. And words have no direct efficacy to alter things; their power is indirect, through their effect on men.

In some forms of magic, symbols were thought to have power over things, but often their power was only over intelligences, spirits, or demons who accomplished the ends desired. Faust's power lay in his temporary control of Mephistopheles. On the other hand, when words have direct control over things, either those "things" are animate and intelligent or the "words" are mere uttered sounds—not symbols at all—functioning as natural phenomena, perhaps as vibrations in the air. When Ali Baba opened the door of the thieves' cave by pronouncing "Open Sesame," that was magic, and either the door is "animate" or the sounds he uttered would have to activate the mechanism the way the beam does a photoelectric cell.

*A* is a natural sign of *B* if *A* is evidence that *B* will occur. But a natural sign need not be a natural phenomenon; it may be an artifact. Rails are as much a natural sign of the existence of a train as dark clouds are of rain. Even symbols may be treated as natural signs, in so far as their *existence* or *use,* not just their meaning, is evidence of something. If a language has twelve words which are synonyms for "house" and no words for "home," we may infer several things about the culture which produced that language. When a man makes a mistake in speech he may be revealing something about himself that would not have been shown by the correct statement. The psychiatrist A. A. Brill reported an incident that amused Sigmund Freud. At a dinner to help launch Theodore Roosevelt's candidacy for president and to honor his "square deal," an ungenerous host served only sandwiches and lemonade. One of the hungry guests said to the host, "You may say what you please about Teddy, but there is one thing—he can always be relied on; he always gives you a square meal."

Symbols, we have said, require an interpreter who knows something of the system by virtue of which the symbol has meaning. Natural signs require an interpreter who knows a theory, or general law, in terms of which the sign is an index of the existence of something else. To interpret dark clouds to mean rain or red litmus paper to mean an acid solution, one must know something general about the connection between clouds and rainfall or litmus and acid. Without symbols there are no scientific laws, and without scientific laws there are no natural signs. So there is a dependence even of natural signs on language.

Because symbols are arbitrary, in that any symbol we agree on will carry the meaning we give it, we may be tempted to infer that no symbol is a better sign than any other. But we should be wrong. It makes no difference that we use the word "horse" for a particular beast of burden instead of any other word we could invent, but it made an enormous difference to the development of mathematics that Roman numerals were replaced by Arabic. Simplicity and flexibility are virtues in a system of symbols. Regular prefixes and suffixes which can be used with many words make it easier to create and understand new meanings.

Although most people in our own culture know that words are arbitrary and are not parts of the things they signify, they often belie this knowledge by their behavior. It is expected that the act of love will be described only in technical terms in the classroom, the drawing room, and the lecture hall. So too with excretion. And death is often described in euphemism: "He passed away." It is as if Anglo-Saxon words are part of the thing described, while Latin roots preserve a distance.

There are three chief ways of using words: to express the feelings of their user, to arouse a response in the hearer or reader, and to stand for a referent. These are the expressive, evocative, and referential functions of language. Much of the time words are used in all three ways at once, but usually one or another function is emphasized. Expressive and evocative elements are almost always present, and when there is reference as well, it affects expression and evocation. Exclamations are almost exclusively expressive. "Damn!" or "Oh, God!" may be uttered even when one is alone, and are clearly not intended to have a referent. Commands are basically evocative. So are sales talks, which are aimed at a response—purchase—and which describe a commodity only in order to stimulate the response. Directions given to a stranger are referential and are quite neutral as expression or evocation. We do not care whether he goes to this or another destination, but we try to use words to describe a route accurately. Scientific writing at its best strongly emphasizes the referential, whereas literary art is a combination of the expressive, evocative, and referential, and neglects any one of these functions at its peril.

Only when language is referential can its meaning be gathered by attending to language alone. When language is expressive or evocative its meaning depends in good part on the circumstances in which it is uttered. "Look out" is usually a warning of danger, but the words alone do not convey the full meaning of the speaker, which may be "duck," "jump," or "stand still." It all depends on the kind of danger confronting the hearer or the speaker's interpretation of that danger.

Sometimes the word "symbolism" is used in a special way, to describe a kind of art in which the key symbols have not only their ordinary referents but also a larger significance. In the first canto of the *Inferno,* for example, Dante speaks of his encounter with a leopard, a lion, and a she-wolf. The names of these beasts refer to the beasts themselves, but they also signify pleasure, ambition, and avarice. Thus they are "symbols" in a special sense, not just as all words are symbols, but in having a secondary, indirect reference for which the interpreter needs additional knowledge. Perhaps the basic point about this use of "symbol" is that the word has its ordinary single referent and that referent in turn is thought of as a symbol with still another referent. Dante (actually his translator) uses the word "lion" to stand for the lion, an animal, but the beast himself is a symbol of ambition. This is not ambiguity, for the word does not have two unrelated meanings; it means ambition *because* it means lion.[5] It is such a chain of reference which perhaps characterizes the important social symbols, like the flag.

[5] There is a current literary use of "symbol" which should be noted. A symbol in a poem may be thought of as in great part referring to other symbols in the same poem instead of to anything outside the poem, although of course the meaning of each symbol, like that of any word, comes from outside the poem. This is, in its internal relations, like mathematical symbols which refer to other mathematical symbols rather than to the external world, or like musical phrases whose meaning is found within the composition itself.

There is still another technical use of "symbol" that is important. Traditionally, art historians have used "symbol" to mean particular emblems, like the thunderbolts of Jupiter or the wheel of Catherine, which constitute a sort of code in which thunderbolts or wheels mean Jupiter or Catherine.

But this is only one point about a complicated matter which is still obscure.

Action, too, can be symbolic. In so far as an act accomplishes something directly, brings about an end, it is functional. But many acts are performed for the sake of their meanings—whether or not the acts affect anything materially—and so are essentially symbolic. Ritual is symbolic in that it is an expression of meaning, although it may also be thought to bring about specific ends. Ceremony and etiquette are symbolic: the hat is removed or the knee bent as a sign of respect. The individual to whom the ceremony is addressed is usually thought of in his status, not his person. Soldiers are taught that they salute the rank, not the man; and it is the Queen, not Elizabeth, to whom the lady at court curtseys. Etiquette is a symbolism of rank and prestige, as well as a way of regulating social intercourse.

Symbols, then, may be words, things, or actions. They are conventional signs which can be interpreted properly only by knowing their socially determined meanings. And those meanings are not decided explicitly at conventions of lexicographers. They develop out of the needs and uses of each society. They reflect the experiences of the people by the objects and relations they symbolize, and the values of the culture by the emphases they provide.

Symbols often have meanings in addition to their ordinary ones which do not appear on the level of consciousness but emerge in dream and fantasy. Under hypnosis or the influence of truth serums people do not merely reveal things they previously refused to divulge; they also reveal things of which they were not consciously aware. A hypnotized patient may remember where he lost something. In addition, things become symbols, take on meanings not consciously attributed to them. Some of these symbols, like snakes and birds, are so widespread that a number of people believe them to be universal among men, not limited to particular cultures. It still seems unlikely, however, that isolated peoples who live in climates with neither snakes nor birds would use them as symbols of any kind, conscious or unconscious. Within our own culture, in any event, there have been some provocative experiments on communication of dream symbols. Farber and Fisher, for example, worked with people under hypnosis.[6]

[6] Described by Dr. Lawrence S. Kubie, "Communication Between Sane and Insane: Hypnosis," *Cybernetics,* Transactions of the Eighth Conference, Josiah Macy, Jr., Foundation, 1952, p. 96.

. . . They described to Subject *A* under hypnosis a painful "experience," and then asked him to dream about this implanted "experience." The subject thereupon reported a dreamlike, disguised reproduction of the "experience," transposed into more or less classical symbols, almost as if this naïve individual had read a dictionary of dream symbolism. Then the experimenters showed Subject *A*'s hypnotic dream to Subject *B* while under hypnosis, and asked Subject *B* under hypnosis what the dream of Subject *A* had meant. Thereupon Subject *B* translated the dream of Subject *A* back into the essence of the unpleasant story which had originally been told to Subject *A*.

That is more than a trick. It is a very important experiment which has been repeated often enough to make us sure that one human being can on a dissociated level translate accurately the symbolic representations of the unconscious content of thought and feeling of another human being, thus exposing and uncovering repressed experiences which the other has had in the past. This is a critical and conclusive demonstration of the power and specificity of unconscious psychological processes, and of their importance in the communications which pass between men.

Our responses to the symbols of nation, family, religion, and love—to mention a few of the things most important to our lives—probably have an unconscious as well as a conscious element. Behavior in response to the use of these symbols is often more extreme than their conscious meanings would warrant. We may risk or even sacrifice our lives when a purely "rational" attitude would lead us to neither. "To give one's life for a belief," wrote Anatole France, "is a high price to pay for a conjecture." The translation of "belief" into "conjecture" retains much of the conscious and referential meaning of the word "belief," but it loses the deep emotions it evokes and the layers of meaning. Such translation in terms of reference alone is necessary for science and exact thought, but insufficient for the social function of a symbol, that of welding individual men together to make a community. Social symbols bind us to one another in society and allow us to share the attitudes, emotions, and values that make social groups cohesive. Such symbols may be objects like the flag, the cross or crescent, the eagle, the hammer and sickle; or they may be formalized action of the sort we call ritual.

The wealth of meaning that is associated with the symbol of the cross is more like the ambiguous meanings of an artistic symbol than the single use of a sign in science. In ancient Rome the cross was an instrument of torture and death; that meaning lingers as we associate the cross with the passion of Christ. Christ's passion generalized means suffering, moral goodness, and redemption, and the cross stands for these, too, as well as for Christendom itself, which presumably accepts those meanings. When we place a cross atop a building we may intend only to mark the building as a church, yet all the associated meanings may be evoked at once in the spectator.

Ritual action has the same richness of ambiguity and evokes intensity of emotion in the same way as objects that are social symbols. The ceremony of the mass—which has, of course, specific religious meanings—and the salute to the flag permit us to share the deep emotional commitments which are the cement of society. Secular rituals share with religious ceremony the elements of celebration, dedication, and consecration, which we will discuss in connection with religion. It is not so much the precise content of the rituals or the theoretical meaning we might assign to them that works this spell, but a generalized and often vague feeling that accompanies them. We may, in a given ritual, dedicate ourselves to the service of our country, and the idea of what we are doing may move us emotionally, but the familiarity of the ritual, the others who perform it at the same time, the feeling that we are connected with something larger than ourselves, that we are accepted, that we have a place where we belong—these concomitants of ritual are in some ways the most important things about them.

It is reported that a schoolteacher, listening attentively to a child recite the "pledge of allegiance," heard the familiar words in slightly altered form: "I pledge my legions to the flag, and the Republicans for which it stands; one nation invisible, with liberals and justice for all." Obviously the children who recite this garbled pledge have no clear idea of what they are saying. Yet the emotional charge, what we just called the concomitants of ritual, is probably not impaired—or at most impaired only slightly—and the purpose of the pledge is fairly well realized.

Various traditions in the Western world would have to treat what has just been said as essentially irrational. The kind of thinking that is involved in classical economics and in much of utilitarianism is, for the most part, a calculation of advantage. Advantage, in these intellectual frameworks, is individual, and what is "reasonable" is the attainment of individual advantage with the maximum of efficiency. Of course, it was also believed by classical economists that what is economically good for one will benefit all, and by utilitarians that the ultimate good is that of the greatest number. But the calculation of advantage for one or for the greatest number was still in terms of dollars and cents or quantities of pleasure. These doctrines are so much a part of us that young people, in periods of adolescent rebellion against the conventions among which they live, often regard the ordinary symbolism and ritual of their society as benighted. So they ask why we continue such forms as the funeral service and the marriage ceremony, implying that they are "useless," a word which in the tradition of utilitarianism is virtually a synonym of "irrational."

After all, why should we not bury the dead quickly and hygienically, marry and divorce merely by recording our intentions, try to live as "rationally" as possible? The answer is that, apart from their purely religious significance, funeral rites and symbols reaffirm each man's place in society, in the family, in friendship, and in love. They commemorate his existence, the particulars of his life, and the gap he leaves in the lives of others. A man is not a machine and is not replaceable in every detail by any other man; each man is unique and this uniqueness, too, is commemorated. Society expresses in the funeral its interest in each of its members and its continuance in its surviving members. It affirms both the death of a man and his continuance, in memory and influence, in his family, his friends, his society itself.

Classical epic and tragedy offer vivid examples of the social importance of one man's death. Homer's *Iliad* ends with the funeral rites of Hector, leader of the Trojan forces, heir to the throne. Why did Homer not conclude his epic with Hector's death? Because the funeral exhibits the meaning of that death to Troy, to Greece, to future ages. Hector was the great champion of his people. While he lived, Troy survived. His death meant the destruction of Troy, the end of all her greatness, the return of the Greek kings to their own lands, even a new direction for civilization.

*Hamlet,* too, does not close with the death of the Prince. There is a symbolic soldier's funeral and an oration by Hamlet's successor, Fortinbras. The meaning of Hamlet's life is stated and his virtues extolled. It is not just *a* prince who died, but *this* prince, and his death deprived his country of his gifts—rich gifts, for he had been "the glass of fashion and the mould of form," and "was likely, had he been put on [the throne], to have proved most royally."

The protagonist of classical tragedy was the king, the prince, the hero. His death was important to his community as a whole. But the life of even the humblest member of society makes a difference to some few others and so, in a way, to his society. By commemorating his life and his death in appropriate rite and symbol, we give "meaning" to our own lives and deaths, an importance that is never for ourselves alone. The "meaning" of a man's life and of his death is not only in himself, but also in his relations to others as individuals, and in his place and role in society. As John Donne said, "No man is an *Iland,* intire of itselfe; every man is a peece of the *Continent,* a part of the *maine;* if a *Clod* bee washed away by the *Sea, Europe* is the lesse, as well as if a *Promontorie* were, as well as if a *Mannor* of thy *friends* or of *thine owne* were; any mans *death* diminishes *me,* because I am involved in *Mankinde;* And therefore never send to know for whom the *bell* tolls; It tolls for *thee.*"[7]

Is it "reasonable," then, to eliminate the funeral service? Obviously not. The calculation of our own advantage that would impel us to dispose of the human dead like so much waste matter would yield, on a larger and *more reasonable* calculation, great disadvantage. It would impoverish our lives and destroy much of their significance. It is not "reasonable" to judge every action by its consequence for direct personal advantage; nor is it "reasonable" to approve ritual and symbolic action only because *enlightened* self-interest finds greater advantage in preserving society than in attaining personal ends in opposition to society. If we act out of enlightened self-interest alone we will have destroyed the emotional and traditional base of society.[8]

For some intellectual purposes it is useful to treat every human action as a choice or decision based on a calculation of advantage. The study of economics has profited by such treatment. But we should never forget that this is only a way of stating things that helps us perform particular intellectual tasks; it is not an adequate description of the nature of man. To be sure, man may find living in society advantageous, but the man who comes to that conclusion is one who has been raised in society, not in the "state of nature." Any man may prefer one type of society to another, but his judgment is already condi-

[7] "Devotion XVII."

[8] Of course, when rites are performed by rote or cheapened by extraneous sentiment, they become shoddy and hollow. After witnessing such rites, a spectator may turn from them in disgust and conclude that all rites—for he may see no better ones—are mere superstition.

tioned by the society that has made him what he is. Santayana says wisely, "It is in the subsoil of uniformity, of tradition, of dire necessity that human welfare is rooted, together with wisdom and unaffected art, and the flowers of culture that do not draw their sap from that soil are only paper flowers."

The bonds of society are the shared symbols, rituals, values, and beliefs of its members, and it is in these that the "meaning" of the society is contained. Bonds may restrain, like the chains on a slave, or they may sustain, like the climber's rope. Social bonds do both. They contain us within the limits of social approval, yet they provide the ways in which we develop our lives and our cultures. There is an analogy here to the forms of art. The sonnet limits its author severely, yet it can give his work incisiveness and power. Personal freedom yields little of value outside a social order and tradition; it can in fact be guaranteed only in a political and social order, and the value of individual differences, of choice and free intelligence, and the arts and sciences which they nurture, are dependent on shared ritual, symbol, and emotion. Our social roots are often forgotten because they are firmly embedded and raise no practical problem. If we forget them overlong, we may uproot them, and with them, ourselves.

# 10

# Religion

The most vital symbols and the most persistent and profound rituals have been religious. Our oldest values and traditions are rooted in religion. They are fresh and strong as long as religion is. They decay with the decay of religious belief. The overwhelming problem of the twentieth century may be that of maintaining common values, indeed of holding society together, in the absence of a broad and firm religious base.

To the devout religious mind the universe is man's home, right is eternal, and the agonies of spirit are solaced by the finalities of faith. For the unbeliever, the difficulties of avoiding sheer nihilism can be enormous. He must justify his values without recourse to God's plan, he must find enough people who hold the same values to preserve a community, he must keep or create rituals which bind men to each other without divine sanction. And he must find a place for man in nature, which easily seems—without religion—alien, inexorable, and valueless. His simplest recourse is not to think at all, but to accept the blind conditioning of his childhood. Or he can accept secular powers as though they were religious: he can divinize the state, glorify its rulers, subordinate his person to the unrelenting pressures of society. All other alternatives are intellectually complex. Some men try to find in nature or history the kind of pattern that most men once found in the will of God. Evolution and progress may thus replace teleology and the perfectibility of man. And God Himself is sometimes replaced by nature or society, science or art.

The mysteries of religion are mysteries of communion and communication. We shall deal in a way with both, but to specify our approach, we must first distinguish between religious study and the study of religion. The dispute over teaching religion in the public schools, for example, is a quarrel about religious teachings, not about the study of religion. Creeds are controversial, and learning them is not a matter of science but of scholarship or faith. When religion is studied in Sunday schools, parochial schools, and some public schools, it is as instruction in a particular faith. In consequence, other aspects of religion which can be studied scientifically and without rancor are neglected: the nature of religion as such, its role in society, the relations between different religions, the changes in religion that take place in history, and the changes in history that take place because of religion.

Such issues and aspects of religion are our sole concern in this chapter. The questions whether religion is true, what it would mean to say religion is true, or whether any one religion is the true one, are beyond our scope in this book. Our essential problem is the place of religion in society.

Religion is a universal institution. Every culture we know of has, or had, a religion, although it is only recently that we have recognized this. As late as the nineteenth century, travelers told of primitive peoples who had no religion whatsoever; but it is now generally agreed that these travelers failed to recognize religion in any unfamiliar form. There are, after all, religions without theology and religions with a minimum of myth and legend. And there need be nothing surprising about the great variety of religions; the family, economic activity, and education are a part of every culture too, but they are often very different.

What of the objection raised by some devout people that there is, in fact, only one religion—and that one, of course, their own? The answer is that they are equating the word "religion" with "true religion." They are not denying that there are other "faiths." As they use the word, "religion" is honorific instead of descriptive. And what we mean by "religions" is what they mean by "faiths." In this sense, there is no real objection to the claim that religions exist in all cultures. But does this universality alone entitle us to say that religion is necessary to society? No; the universality of any institution may simply be a remarkable accident. Of course one suspects it is more, but to justify that guess one has to find some basis for the institution in man's life or in his essential needs. Survival and continuance of a community depend on familial, economic, and educational institutions; but it does not seem, at least on the surface, that the same can be said for religion.

Ruth Benedict, recognizing this, commented: [1]

> . . . All other social institutions rise from known bases in animal life . . .
>
> With religion this is not true. We cannot see the basis of religion in animal life, and it is by no means obvious upon which of the specifically human endowments it is built up.

Benedict is perhaps too materialistic in her appraisal of man's needs; there may be needs naturally based on the human psyche rather than on human biology. And what about man's need for the society of his fellows, and so for society? Our basic needs for shelter, food, and sex are satisfied by different societies in different ways, but they *are* satisfied in each society so long as it continues to function. The need for a continuing and relatively stable society is met by education, which transmits the culture, and by religion, which keeps it alive and meaningful.

Separate institutions do not make up a society unless they are treated in thought and conduct as related parts of a whole. Nor do institutions continue unless people prize them. Individual members must be dedicated to their society, accept its ways as more than accident or expediency, cherish its traditions and meaning. They must be comfortable with cultural detail and must find in their society and its institutions a meaning for their own lives. Some of these attitudes are developed by sheer use and wont; but others result from the integrative and binding function of religion. Religion gives society a greater significance for its members than it could have purely as an organization for the satisfaction of material wants, and it gives each life more meaning than it could have as only struggle for survival or eminence. And when men cherish their institutions, they are likely to fulfill the obligations those institutions impose. In other words, men become moral, thus keeping their institutions alive. So religion is responsible for much of the cohesiveness necessary for the acceptance and con-

[1] Ruth Benedict, "Religion," *General Anthropology,* Franz Boas, ed., Heath, 1938.

tinuance of a society and for the feeling that man is more than an isolated animal, that he is part of a larger, more enduring whole in which he lives and from which he derives his own significance and importance.

The ancient historian Polybius found great social virtue in acceptance of religion, although he obviously did not regard it as true!

> The point in which, to my mind, the Roman constitution evinces its greatest superiority is the attitude adopted toward Religion. In my belief, a disposition that is reprobated in other countries is actually the keystone of the Roman system, by which I mean Superstition. At Rome, this feature has been artificially exaggerated and introduced into private life as well as public affairs to the utmost extent conceivable. Doubtless many readers will find this extraordinary, but in my opinion the Romans have done this deliberately with a view to the masses. If a community exclusively composed of intelligent beings were a practical possibility, such a policy might conceivably be unnecessary; but in fact the masses everywhere are unstable and instinct with such antisocial passions as irrational temper and homicidal fury, and there is therefore no means available for holding them in check except unseen terrors and the mummery of Superstition. From this point of view, I feel that there was nothing random or irresponsible in the policy of our forefathers when they introduced among the masses the conceptions of Religion and the notions regarding Hell, and that it is far more irresponsible and irrational of the present generation to expurgate these ideas. One of the many unfortunate consequences of this false step may be seen in the fact that, in Hellenic countries, when even so trifling a sum as one talent of public money is entrusted to persons in official positions, the latter are controlled by ten countersignatures and as many seals and twice that number of witnesses and are still incapable of being faithful to their trust; whereas, in Rome, public men handle vast sums of money in the administration or on diplomatic missions upon the sole security of their personal oath, and are still faithful to their duty. In other countries, it is rare to find an individual who keeps his hands off the public purse and can show a clean record in this respect. In Rome it is equally rare to see anyone convicted of such malpractices.[2]

In the ancient world each people identified itself with its religion as in the modern world we identify with the nation-state. William Robertson Smith, nineteenth-century pioneer in the study of early

[2] Arnold Toynbee, ed., *Greek Historical Thought*, New American Library, 1952, pp. 124 ff.

religions, said that an ancient people or nation could be best defined as one having a particular religion. Every nation was entitled to its own rites and its own gods. When a man changed his national allegiance he quite naturally changed his religion as well. It was not so much that a nation regarded its own religion as better. It was enough that it was theirs. In the great Biblical passage in which the widowed Ruth decides to live with her mother-in-law, Naomi, she says, "Thy people shall be my people, and thy God my God."

Of course, the art and literature of ancient peoples, their public ceremonies, and their special customs set them off from each other as members of individual cultures. Yet all these things were either religious in themselves or were connected in some way with religion. The content of Greek tragedy in the classical period, for example, was religious. Religion, then, permeated the entire culture of a nation and gave it its special character. Our question now must be: What is the nature or quality of religion that it can be so fundamental to culture?

It would be presumptuous to attempt an exact definition of religion. What seems to be the meaning of "religion" as it is used in ordinary conversation today is the worship of God, or of gods. And when we use the adjectival form ("He is a religious man"), we usually mean one or more of several things: that this man is a member of a church, that he believes in the existence of God, that he holds a philosophy, a view of life, that attributes a quality of great value, sacred or approaching sacredness, to some things.

None of these meanings, however, is adequate to the variety of religious expressions and institutions found throughout history. To consider first the belief in God: there are religions which seem to have no God in any sense in which we ordinarily use the word (pure Buddhism, for example), and others in which the meaning of God is, to say the least, strained (as in some kinds of Unitarianism). Even if we define God, or the gods, loosely, we find it difficult to know whether we are talking about something that has the same significance in different religions. There are two ways in which we get some idea of the meaning of the spirits or supernatural powers believed in by people of diverse religions: first, by the content of the belief, what they *say about* those spirits or powers (description, myth, theology); second, by their *behavior toward* them (ceremonies and individual actions). The supernatural beings include ancestors, familiar spirits, a

Supreme Being. Attitudes toward them range from a kind of contempt that does not exclude fear, to utter worship, and sometimes involve willingness to oppose a god and unwillingness to obey him.

Can a spirit be thought of as a god if his power is limited? If we grant the possibility, how far can we go? As we examine spirits of less and less power, at what point do we have to say we are no longer talking of gods? It is not even true that a pantheon of gods share among themselves the power that modern Western religions attribute to one God: The Norse gods, for instance, were limited by Necessity, or Fate, and were doomed to destruction. The omnipotence of God, like His oneness and the exclusive truth of the religion that accepts Him, is a notion introduced by the Jews. It penetrated Western civilization late in its history and was never fully accepted by simple folk, who see God's power as diffused through angels, seraphim, cherubim, and saints, and as opposed—perhaps limited—by Satan and his demons.[3] There can be genuine religion without an omnipotent God.

A god quite different from our ordinary notions of a deity is the Trickster, common to many primitive religions. The Trickster is a spirit who may be mischievous or vindictive, who is not fully to be trusted, and who may even, like the Norse god Loki, play tricks on the other gods. The American anthropologist Paul Radin characterizes one of them as "obscene, a fool, a coward, and utterly lacking in self-control." Primitives may resist or try to outwit their trickster gods, but they clearly do not regard them as demons because they expect them, although capricious, to be kind and generous on occasion.

But can a spirit be classified as a god if his followers do not worship him as we understand "worship"? Perhaps. But the meaning is strained further. There are kinds of worship that are not based on love or fear as our worship is, but contain many elements alien to our religions. Some of these are strangely combined: many of the Melanesians, although fearful, try to outwit, cajole, or bribe their gods. They offer inferior sacrifices, hoping the gods are fooled. Or the god

[3] The ancient gods of Greece and Rome persisted in the popular imagination long after Christianity had conquered the West. But the gods were converted to elves, trolls, goblins, gnomes, fairies, and other spirits. Pan reappears as Puck or Robin Goodfellow. The fairy tales of all Western nations contain the gods of the woods and the household, but under new names. Sometimes the gods are demons, and Satanism is occasionally an appeal from the power of God to that of older gods.

must be worshiped and obeyed even though what he commands is both terrible and ordinarily forbidden. God commanded Abraham to kill his only son, Isaac. But He stopped Abraham in time, just as the knife was raised. In primitive religion the awful act the god commands is not always stopped, nor is it less repugnant because of its divine sanction.

Among the Kwakiutl Indians of Vancouver Island, for example, there is a real horror of cannibalism, yet the eating of human flesh is a divinely ordained ritual. One group is allowed and enjoined to practice it and therefore holds higher rank than any other. The initiation of the Cannibal Dancer includes eating from a prepared corpse and goes beyond the consumption of dead flesh: the frenzied initiate bites chunks from the arms of other celebrants of the rite. As the cannibal dances before the corpse the others sing: [4]

*Now I am about to eat,*
*My face is ghastly pale.*
*I am about to eat what was given me by Cannibal*
*at the North End of the World.*

"Cannibal at the North End of the World" gives the food and presumably demands that it be eaten. But the Cannibal Dancer not only "pales" at the thought; immediately after the ceremony he is isolated for four months, during which he purifies himself of what is regarded as defilement. And when his seclusion is ended and he can rejoin his fellows he is not yet a man among men: for four years he cannot work, gamble, or touch his wife.

So there are religions which conceive their "gods" and practice their worship in ways quite different from our own. Conceptions of deity are of three general types. Modern Western religions are all *monotheistic*. That is, they hold that there is but one God who is the supreme power of the universe. *Polytheism,* which was typical of the ancient world, is a worship of many gods, often thought of as constituting a more or less organized hierarchy of power. Polytheistic religions usually are concerned with the worship of only their own gods, but do not exclude the belief that the gods of other peoples exist and have power. Even the very early Hebraic religion, which already worshiped only one God, did not hold Him to be the single divine power in the

[4] Quoted from Ruth Benedict, *Patterns of Culture,* New York, Mentor, 1946.

universe; the Jews at that time admitted the existence of the gods of other peoples but did not worship them, and they thought of Jehovah as the *true* God, angered by the worship of false gods. This was not polytheism, because only one God was worshiped, nor was it yet fully monotheism, because it was believed that other gods did exist. The name given to it is *henotheism*.

Many ancient peoples believed that their gods aided them in their disputes with others and that the gods of the enemy opposed them. It was common for the peoples of Arabia (identifying an image with the god) to take the god into battle so that the force of the deity would be added to that of his followers. Perhaps it was an interpretation of Judaism (by then monotheistic and spiritual) in terms of their own more material religion that led the Philistines to steal the Ark of the Covenant, as if the God of the Jews resided in it physically. Such misinterpretation is common when religions with differing concepts of God come into contact.

If we cannot define religion as belief in God, can we define it as a philosophy, a set of beliefs? No; that would rule out many early and primitive religions, because they are essentially systems of ritual accompanied by myth. Even religions with systematic theologies could not be defined as religion merely because of the *content* of those theologies. The central beliefs of some are missing in others; redemption and immortality, for instance, are basic Christian doctrines but they are not found in all other theological religions, and one, Buddhism, explicitly opposes the doctrine of individual immortality. Religion, then, must be understood apart from such special concepts.

What else can be used to define religion? Perhaps conduct common to all of them. Religious behavior always includes *taboo* (derived from the Polynesian *tabu*, which means forbidden). We have already considered some things ordinarily thought of as taboo, although we did not use the word: Abraham's sacrifice of Isaac and the Kwakiutl's eating of human flesh. But objects and persons, as well as actions, can be taboo. An action that is taboo is forbidden; a person or thing that is taboo is not to be touched.

Taboo as a religious prohibition is probably connected with the notion that the supernatural has power and can be dangerous: disobey its injunctions and it will injure you. At the very least, if one touches an object that is taboo, he himself becomes taboo and must then undergo ritual purification. (In our own religions there are ritual prohibitions; the Roman Catholic must not, without dispensation, eat meat on Friday, nor may the Jew or Mohammedan, without special permission, eat the flesh of the pig.) However, it is possible to violate taboo accidentally: if a Polynesian accidentally touches a corpse or if a tribesman of the Kikuyu is splattered by a kite or a crow, a taboo has been broken. Here, too, there is usually a change in ritual status which can be remedied, if at all, only by purification. Not all taboo objects and persons are ritually unclean, however; some are too sacred to be touched. Polynesian chiefs, for example, are holy; they may not feed themselves, but are fed on long prongs which are handled with great dexterity to avoid touching the sacred teeth.

Although some taboos exist in all religions, taboo is inadequate as a definition of religion because it does not distinguish religion from other things: there are social acts of the same kind. On formal occasions it is "taboo" for an ordinary mortal to turn his back on a king, the Pope, or the Prime Minister of Canada. It is also "taboo" for a man to wear his hat in an elevator when a lady is present (unless the elevator is in an office building where, presumably, "ladies" are mere women).

Yet the *concept* of religious taboo, of actions and things too holy or too unclean to be performed or touched, points to an attitude of which taboo is only an example, to a special way of regarding and treating objects which may be central to all religion.

One of the most fruitful suggestions about the essence of religion came separately from two great sociologists, Emile Durkheim and Max Weber, who made a distinction between the ordinary, secular, or mundane, and the special, sacred, or holy, and who argued that religion includes the acceptance of some particular things as sacred. Weber used the word *charisma* to stand for this quality of sacredness, of being "set apart" from the ordinary. (*Charismata* in New Testament Greek means literally "things freely given," and is usually interpreted by theologians as what comes from God's grace.)[5]

[5] Durkheim's distinction was stated as that between the "sacred" and the "profane," and Weber's as that between the "charismatic" and the "ordinary." Robertson Smith had already made a similar distinction between the "holy" and the "common." These three sets of terms are not identical, although for some purposes they can be treated as such. As Durkheim used "sacred" it included the ritually unclean (as in religious taboo) and what Smith called "holy. It is not common in English to think of the sacred

Although the notion of the sacred or charismatic is extremely helpful in thinking about religion, it is not easy to characterize religion in this way only, because there is a feeling of sacredness about some things that are not ordinarily thought of as religious. Weber himself used the word *charisma* for the magnetic qualities of a political leader; it could be used equally well in talking of flags and other national symbols. Medieval chivalry treated the lady to whom the knight's service was dedicated as in some sense sacred, and romantic poets have written of the beloved in the same fashion. So although a feeling that particular objects are sacred may be characteristic of all religions, it is not sufficient in itself to distinguish religion from other activities and institutions of man.

To make this distinction, perhaps we must add to the sacred two other elements of all religions. The first is a recognition of *power* beyond ourselves. It need not be God's, but it is a power that we do not ordinarily possess, that comes to us, if at all, in moments of inspiration. Thus the power may be external, and may enter us only when certain conditions are fulfilled. The early religion of the Romans revolved around a belief in what they called *numen,* a kind of power that existed everywhere and expressed itself in human action. The old tradition persisted even in later Rome so that a general, celebrating his triumph after successes in the field, led a procession to the temple of Jupiter and there offered up "the achievements of Jupiter wrought through the Roman people." By this time Jupiter and the other gods and goddesses were personifications of aspects of *numen,* which was not originally associated with a god or spirit.

The other necessary element in religion, in many ways the basic one, is *ritual,* the prescribed and formal acting out of a ceremony, usually repeated in exactly the same way on specified occasions. A

---

as including both the holy and the unclean, although the Latin *sacer* does mean both holy and accursed (i.e., dedicated to either supernal or infernal deities). Weber, by his use of *charisma,* emphasized the special, set-apart quality of sacred objects and the nonutilitarian basis of that quality: some objects command devotion just because they are endowed with *charisma,* not because of any profit to be gained in their service.

Note that we are using these distinctions for our own purposes. Originally, they had various contexts. Weber, for example, thought of *charisma* as an element in a general "primitive religion." And charismatic objects included "soul" and "gods." But *charisma* also distinguished the supernatural aspect of things and actions from their ordinary or secular aspect.

system of belief, even one that has a place for God, is a philosophy, not a religion, unless it explains, or is expressed, in a ritual.[6] But a ritual without a system of belief may still be a religion. And it is in ritual—especially in ritual as symbolic—that much of the social function of religion can be found.

It is highly probable that early religions consisted almost entirely of rites and that both myth and theology developed later. Myth may, in the first instance, have been a *symbolization* of rite,[7] and theology may have been an *explanation* of both ritual and myth. Of course, one can cite the emergence of a religion like Protestantism as in part the result of doctrinal difference and theological dispute; indeed, the same is true of Christianity itself. But both of these religious movements developed out of prior religions which had, by that time, a fully evolved doctrine. These are cases of nature imitating art. As an analogy: political theory was developed fully only after there were political institutions. One may think of the Constitution of the United States as a social contract to which the individual states were parties, but it does not follow that the origin of society was in a contract among men. Yet few followers would be drawn to a new religion or a new polity today which was not based on explicit principles, despite the fact that religions and states did not come into existence originally in so rational a manner.

One can scarcely do better than to quote William Robertson Smith on the points we have been making: [8]

> . . . We shall go very far wrong if we take it for granted that what is the most important and prominent side of religion to us was equally important in . . . ancient society. . . . In connection with every religion, whether ancient or modern, we find on the one hand certain beliefs, and

[6] Sunday supplements recurrently feature articles with such titles as "What Does a Methodist Believe?" or "What Does a Baptist Believe?" in the conviction that this is the heart of the matter. But there are no articles about what a Methodist does or what his observances are, probably because these questions are thought to be of importance only to Methodists (who learn the answers from their clergy) but of no concern for the general reader, who just wants to know what Methodism *is*.

[7] Myths, being stories, do not literally state or describe rites, but contain symbols and allegories which imply them.

[8] William Robertson Smith, *Lectures on the Religion of the Semites,* Adam and Charles Black, London, 1894, pp. 16 and 17.

on the other certain institutions, ritual practices, and rules of conduct. Our modern habit is to look at religion from the side of belief rather than of practice; for, down to comparatively recent times, almost the only forms of religion seriously studied in Europe have been those of the various Christian Churches, and all parts of Christendom are agreed that ritual is important only in connection with its interpretation. Thus the study of religion has meant mainly the study of Christian beliefs, and instruction in religion has habitually begun with the creed, religious duties being presented to the learner as flowing from the dogmatic truths he is taught to accept. All this seems to us so much a matter of course that, when we approach some strange or antique religion, we naturally assume that here also our first business is to search for a creed, and find in it the key to ritual and practice. But the antique religions had for the most part no creed; they consisted entirely of institutions and practices. No doubt men will not habitually follow certain practices without attaching a meaning to them; but as a rule we find that while the practice was rigorously fixed, the meaning attached to it was extremely vague, and the same rite was explained by different people in different ways, without any question of orthodoxy or heterodoxy arising in consequence. In ancient Greece, for example, certain things were done at a temple, and people were agreed that it would be impious not to do them. But if you had asked why they were done, you would probably have had several mutually contradictory explanations from different persons, and no one would have thought it a matter of the least religious importance which of these you chose to adopt. Indeed, the explanations offered would not have been of a kind to stir any strong feeling; for in most cases they would have been merely different stories as to the circumstances under which the rite first came to be established, by the command or by the direct example of the god. The rite, in short, was connected not with a dogma but with a myth.

Rituals are dramatic and symbolic actions, as myths are symbolic beliefs. In ritual something of significance to the communicants is performed, like an act in a play: sometimes the event is imitated directly; sometimes it is symbolized by a dramatic action that is not imitative; sometimes it is a little of both. In the Eucharist (the central rite of many Christian churches), for instance, the communicant may take in his mouth both the wafer and the wine, or a wafer dipped in wine, and this dramatic action both symbolizes and in a limited way imitates partaking of the body and blood of Christ.[9]

[9] In some churches it is interpreted as literally partaking of the body of Christ, which actually is the Host.

There are striking uniformities in the things dramatized in the rituals of different religions. An obvious one is implied by the seasonal character of much ritual. There are many rites, for example, connected with the sowing of seed and the reaping of harvests. These are usually dramatized in rituals of death and birth; the god dies when the grain is harvested and is reborn in spring. A common variation combines the death and resurrection of the god in a single ritual or in a series of rituals performed within a brief span.

Spring festivals, however, are not concerned only with grain and vegetation: the sap stirs in the veins as well as in the trees. So rites of earth's fertility may also be rites of love and puberty. And the two are one. The mystery of creation or generation is often celebrated in a sexual ritual because sex is the obvious symbol of new creation or birth. Some early fertility rites culminated in the public copulation of a man and woman; and there are puberty rites in primitive religions in which the girl is made a woman sexually by one of the older men. When human fertility and fertility of the soil are celebrated together, man and nature are related. And when coming of age sexually and coming of age socially and politically are dramatized in a single ritual, man and society are related.

Still other uniformities are apparent. In one ritual form or another, religions dramatize birth, marriage, and death. These different ritual forms are analogous to the different institutional forms in which each society organizes its economic, familial, and educational life. Each particular form expresses the attitudes, sentiments, and evaluations of one group as it focuses on the universal subject matter of the rite or institution. Man is not taught by words alone: continuance of the particular rite, like that of the institution, is part of the transmission of a culture.

Yet, whatever the form of the ritual, the threefold relationship of man, nature, and society appears again and again: an event in man's life is connected with the course of nature and is expressed as a link in the chain of society. *So ritual celebrates the major events in human life, accords them social recognition, and relates man and society to nature.*

In part, ritual drama expresses what already exists: man's natural being and his social behavior. In part, ritual transforms the sheerly individual into the social. The puberty rite, for instance, does not merely dramatize an event that would take place without it, the initiation of the youth into the group. It *is* that initiation. The youth becomes an adult member of the group because he is ritually inducted. Later, when he is a celebrant at the induction of others, he plays the part of a member, thus affirming the status that once was accorded him.

Of course, many social rituals are not religious at all and others are religious only in part. Etiquette, the patterns of courtship, and the formal conduct of tribunals and legislatures are nonreligious rituals, whereas the coronation of kings and acceptance of public office are semireligious: kings are consecrated by church dignitaries and the oath of office is usually sworn on the Bible. The distinction between purely social ritual and religious ritual is that the latter, but not the former, includes reference to a more than human power and more often centers around charismatic objects.

Even now when other institutions have encroached on what was once the domain of religion, social ritual preserves some of the drama of religious ritual and performs some of its antique function. In the marriage service, whether religious or civil, society takes cognizance of individual choice and affection. The lovers are no longer just two people finding joy in each other but are transformed into a family. The birth of a child is not only the fruition of human love but the addition of a member to society, and it is recognized as such in the ceremonies following birth.

But ritual does not dramatize only events in the lives of individual men. It often commemorates historic events in the life of the group or in the development of its religion. The Jews, for example, celebrate their delivery from bondage in Egypt and the making, through Abraham, of a covenant with God. These historic incidents are commemorated, kept alive in memory, as traditions that foster the continuity and uniqueness of the group. After all, a group, like a man, is what it is in great part because of its remembered history.

Ritual, then, chiefly expresses *celebration, consecration, dedication,* and *commemoration.* In the ritual drama of puberty the youth is *dedicated* to his social role and *consecrated* to his religious function. His sexual maturity and initiation into the group are *celebrated* together. And the actions of past initiates are *commemorated.* Particular rites may concentrate on one of these, but rarely to the total exclusion of the others. Celebration alone might be merely an expression of joy in the event, but dedication, consecration, and commemoration place the event in a larger context, make joy more than animal exuberance, and affirm shared purposes of greater significance than any private ends.

In the ancient Chinese *Book of Rites* (the *Li Chi*) there is a compact expression of the social function of ritual: "Ceremonies are the bond that holds the multitudes together, and if the bond be removed, those multitudes fall into confusion."

Myths, strictly speaking, are traditional stories about the gods or other supernatural powers. They are distinguished from legends, which are about human heroes, and from fables, which deal with gods, men, and animals in a manner intended to instruct or entertain. Yet most ancient tales combine elements of the mythic, the legendary, and the fabulous. In these tales men have dealings with the gods which may be both edifying and entertaining. The name we give such tales—myth, legend, or fable—depends on their emphases or, if they are didactic, on the morals to be drawn from them.

Myths are "symbolic" because they are not (like theology) literal, expository, and systematic, but mean something, perhaps equivocal, that they do not explicitly state. Even the gods are symbols and, like the myths themselves, do not have a fixed, unequivocal meaning. Pallas Athena, for example, was the goddess of wisdom and as such could symbolize wisdom in a myth. But she was also the guardian of cities, the goddess of Athens, the protector of civilized life, patron of the arts of peace and war, and ruler of storms. In each of the myths about her she might stand for any or all of these, while on the literal level of meaning she was always her divine self. According to Greek myth, Athena had no mother; she came full blown from the head of Zeus. Does this mean that wisdom is a masculine faculty or just that it is the result of thought? In the light of Athena's other functions, does the myth mean that the city is the fruit of civilization and that it is brought about by wisdom? Or does it mean that wisdom is the flower and the city the stem?

Interpreting a myth is always difficult, but we make it even more difficult if we assume that each myth has only a single meaning or that a god symbolizes only one of his varied functions each time he appears in a myth. Myth is, among other things, a form of literary art and it bears the unmistakable stamp of all art: its meanings are ambiguous. In great art there are usually several levels of meaning, and sometimes each of them is joined in a single work to make a larger meaning.[10] So the things Athena symbolizes may be combined as a single idea in which the aspects of the goddess, like those of the Godhead, are separate but related to make a congruous whole.[11]

In ancient religions there were commonly several versions of the same myth, and the worshiper could choose the one he liked best. Indeed, belief in myth was not obligatory—although the performance of ritual was. There is a tendency today to regard ritual as an acting out of myth. But as Robertson Smith said, ". . . it may be affirmed with confidence that in almost every case the myth was derived from the ritual, and not the ritual from the myth; for the ritual was fixed and the myth was variable. . . ."[12]

It is also common today, in comparing modern with ancient civilization or Western with primitive culture, to treat myths as prerational explanations of the universe. And superficial examination does make it seem that *some* myths are allegories of natural process, although others are clearly moral allegories or explanations of ritual. But it may be ignorance of the rituals with which those myths were once connected that leads us to believe that *any* myths are attempts to explain nature.[13] We should not assume that ancient and primitive peoples ask the same questions about the world in their religions that we ask in our science. Their questions may only be about the meaning of ritual, and their mythic answers only explanations of the rites and of their moral and spiritual significance.

It can be argued, for instance, that myths of creation were not explanations of the origin of the universe at all but were literary translations of ritual. The ritual was performed at regular intervals. The myth, in each of its variations, articulated a single meaning for all the performances. There was a Babylonian creation myth in which the god Marduk battled with a dragon; Marduk cut the dragon in half, thus creating the heavens and the earth. Sir James G. Frazer,

[10] For a discussion of meaning and ambiguity in art, see Chap. 11.

[11] In the Middle Ages the levels of meaning in Scripture were formalized as four (called four senses): literal (or historical), allegorical, moral, and anagogical (spiritual or mystical). Dante used all four in *The Divine Comedy*.

[12] Smith, *Lectures on the Religion of the Semites*, p. 18.

[13] Myths were often written, and so survived. Ritual was usually transmitted by apprenticeship and practice. Wars and migrations might readily blot out the latter but preserve the former.

thinking of myths such as this as primitive attempts at science, called them "quaint fancies." The distinguished classicist F. M. Cornford replied trenchantly some years later: [14]

> . . . The contents of the Creation myth are not "quaint fancies," or baseless speculations; nor are they derived from the observation of natural phenomena. Starting from the given appearance of the starry sky above our heads and the broad earth at our feet, no one but a lunatic under the influence of hashish could ever arrive at the theory that they were originally formed by splitting the body of a dragon in half. But suppose you start with a ritual drama, in which the powers of evil and disorder, represented by a priestly actor with a dragon's mask, are overcome by the divine king, as part of a magical regeneration of the natural and social order. Then you may compose a hymn, in which this act is magnified, with every circumstance of splendour and horror, as a terrific battle between the king of the gods and the dragon of the deep. And you will recite this hymn, every time the ritual drama is performed, to reinforce its efficacy with all the majesty of the superhuman precedent.
>
> Now so long as the myth remains part of a living ritual, its symbolic meaning is clear. But when the ritual has fallen into disuse, the myth may survive for many centuries. The action will now appear crude, grotesque, monstrous; and yet a poet may instinctively feel that the story is still charged with significance, however obscure, owing to the intense emotions that went to its making when it was part of vitally important religious action. Symbols like the dragon still haunt the dreams of our most civilised contemporaries.

Ritual may not always have preceded myth. The two may have grown together or myth, in some cases, have appeared first. The important point is that ancient and primitive religions in their early forms were essentially a matter of ritual, not of belief in myth. But myth was often a part of ritual (recited by the priest) as well as a symbolic explanation of it.

Myth performs the same functions as ritual, although in a different way. As ritual dramatizes the relation of man to nature and

[14] F. M. Cornford, "A Ritual Basis for Hesiod's *Theogony*," *The Unwritten Philosophy and Other Essays*, Cambridge Univ. Press, Cambridge, 1950. A considerable expansion of this argument may be found in another of Cornford's books, *Principium Sapientiae*, Cambridge Univ. Press, Cambridge, 1952, Part II.

society, so myth symbolizes, or allegorizes, those relations. There is also a difference in emphasis. Ritual emphasizes the relation of man to society. Myth emphasizes the relation of man to nature.

However else we distinguish religion, magic, and science from each other, one important difference is obvious. Science includes studies of religion and magic, but neither religion nor magic investigates science. Some nineteenth-century assumptions about society resulted in mistakes which carried the weighty imprimatur of science, and so are still with us.

One assumption, already noted, was that doctrine is the essence of religion—as though all religions were like ours. Social scientists were as guilty of this provincialism as anyone else. And they committed two unique but related errors. If religion is essentially a doctrine, or set of beliefs, then what, they asked, are the beliefs about? Nature. Then religious beliefs are prerational or prescientific answers, they concluded, to questions which can be answered properly only by science. How is it that these answers were made, and why have they not disappeared? Religion, these scientists said, is a stage (perhaps a necessary one) in human evolution. Before man was rational, at least before he had the methods and concepts of science at his disposal, he dispelled his fears and satisfied his curiosity by believing in the supernatural. Although man now has science, religion lingers on, like the appendix or wisdom teeth, as a vestige of an earlier time when it was needed.

Now, the scientific study of religion (its history, its varieties, its role in society) has been enormously useful, and many of the things we have said earlier are conclusions of scientific investigation. But it is misleading to read the attitudes and concepts of science into religion. This distinction hinges on the difference between science as a method of inquiry and science as a cultural force. The former yields knowledge. The latter may habituate people to particular attitudes and limit the range of their imagination and sympathy.

Although science is a product of culture, it also transcends culture and in a way creates a special culture of its own. Its truths are universal, not culturally relative. Science tests hypotheses about cultures and, for that matter, about science. But to accept science as a cultural norm on the basis of which all other human activities are interpreted and judged is not science. It may be called "scientism." [15] And it commits us to all the errors inherent in trying to explain one culture in terms of another. Then all nonscientific thought may be understood (really misunderstood) as a form of science, and so all thought becomes good science and bad science or, if we think in evolutionary terms, scientific and prescientific.

Scientism underlies those errors—oversimplifications, really—in the study of religion which we have spoken of as coming from scientists. One, the rationalist interpretation of religion, is associated historically with the names of Tylor, Frazer, and Max Müller. For the rationalist, religion is a nonscientific, or irrational, attempt to explain the universe. Without the methods of science, its basic empiricism, and the principles of logic, explanation of the universe is essentially mythic or theological. Ancient and primitive myths, therefore, are regarded as alternatives to scientific theories, as explanations of events mistakenly stated in terms of purpose and agency instead of what would be properly scientific, those of cause and effect. Yet the chief type of religious question, we have seen, is about ritual, legend, or myth (sometimes regarded as historical fact), not about the physical universe. The essential concerns of religion are not with the physical universe at all, except in so far as they include the purpose and destiny of the world, matters outside the scope of science.

Another scientistic oversimplification is an evolutionary interpretation which treats religion as a necessary stage in the development of belief. Auguste Comte and Herbert Spencer are probably the best known members of this school. For Comte, religion is succeeded historically in the life of mind by metaphysics, and metaphysics by science, so that religion and metaphysics are not merely intellectual errors but are precursors of science. There are two inadmissible assumptions in this theory of social evolution. The first of these is that evolution has fixed ends. A corollary is the belief that evolution is progressive, that every stage is an advance over the stage before, a step closer to the end. Men think, naturally enough, of their own development from a single cell at the bottom of the sea as "progress"; scientists think of the advent of science in human history as intellectual progress. But what would the unicellular organism think? And what

[15] Here we are concerned with the treatment of religion as though it were science; in the discussion on page 196 we are concerned with the treatment of science as though it were religion.

would be the judgment of a primitive Christian? It is folly to think that whatever comes later is better: Hitler's extermination camps, after all, did not exist until the twentieth century.

The second of Comte's assumptions is that the general stages of evolution are everywhere the same, so that we can predict, on the basis of our own past, the stages through which simpler or more primitive societies must still go. The American anthropologist Morgan and the Swiss anthropologist Bachofen offered presumed evidence from their own field to justify this assumption. Marx and Engels, who accepted both the evidence and the assumption, then were enabled to argue that an older capitalist nation is a picture of the necessary future of a younger one.

Some rationalists did not believe men could be so utterly wrong about physical nature as they thought the priests of early religions were. So they regarded the shaman, the witch doctor, and the medicine man as deliberate frauds who used their office for purely personal advantage, and they explained trances and catatonic states, even mysticism, as deceits practiced on a gullible public. Frazer spread this error widely. William James, who met Frazer in Rome, wrote about him in a letter, "He, after Tylor, is the greatest authority now in England on the religious ideas and superstitions of primitive peoples, and he knows nothing of physical research and thinks that the trances, etc., of savage soothsayers, oracles and the like, are all *feigned!* Verily science is amusing!"

It is difficult from the standpoint of scientism to see the difference between magicians and priests or between magic and religion: they are equally fraudulent and equally erroneous. To make matters more confusing, one element in religion, prayer, seems to all kinds of people to be either a form of magic or like magic. In both prayer and magic something is wanted, or asked for, and the attempt to attain it is by way of a relationship to a supernatural power. Science, then, seems quite different from either prayer or magic. The scientist, it is thought, wants something too, knowledge, but he tries to get it by observation and manipulation of the natural.

But although some religions contain magical elements, religion and magic are not basically alike. One distinction popular among contemporary scholars is that in religion the person is subordinated to the object of worship or external power, while in magic that power is subordinated to the person of the magician and compelled to do his bidding. This is perhaps most clearly seen in the very matter of prayer, in which we may ask a superior power to do something for us, whereas in magic we bend some power to our own will. And, as we shall see, magic is in some ways closer to science than to religion.

Even this distinction between prayer and magic is inadequate because it rests on a superficial understanding of prayer. Although prayer is sometimes a request made of God or the gods to perform a specific service, there are two other important uses of prayer, neglected in most discussion, which seek no specific end: the ritual use and the emotional or poetic use. Prayer as a part of ritual is formal and ceremonial. It suits the occasion, and even when it seems to ask for something it may be just an expression of faith or evaluation, literally asking for nothing at all.

One ritual use of prayer is the thanksgiving or grace spoken before meals. The words are usually uttered in a formal manner with little attention paid to what they say, and in some cases their content is not in any sense a request. Let us compare the words used on this occasion by Catholics, Protestants, and Jews.

CATHOLIC: Bless us, O Lord, and these Thy gifts, which we are about to receive from Thy bounty. Through Christ our Lord. Amen.

PROTESTANT: Bless, O Lord, this food to our use, and us to Thy service, and make us ever mindful of the needs of others, in Jesus' Name. Amen.

JEWISH: Blessed art Thou, O Lord our God, King of the universe, Who bringest forth bread from the earth. Amen.

The general blessing the Catholics ask is scarcely a specific request. The Protestants ask help in giving, not receiving. And the Jews do not even ask the Lord's blessing, but announce that *He* is blessed.

As for the emotional or poetic use of prayer, there is a fine statement by George Santayana: [16]

> . . . A man overcome by passion assumes dramatic attitudes surely not intended to be watched and interpreted; like tears, gestures may touch an observer's heart, but they do not come for that purpose. So the fund of words and phrases latent in the mind flow out under stress of emotion; they flow because they belong to the situation, because they fill out and

[16] *The Life of Reason,* Vol. III, *Reason in Religion,* Scribner's, 1936, p. 39.

complete a perception absorbing the mind; they do not flow primarily to be listened to. The instinct to pray is one of the chief avenues to the deity, and the form prayer takes helps immensely to define the power it is addressed to; indeed, it is in the act of praying that men formulate to themselves what God must be, and tell him at great length what they believe and what they expect of him. The initial forms of prayer are not so absurd as the somewhat rationalised forms of it. Unlike sacrifice, prayer seems to be justified by its essence and to be degraded by the transformations it suffers in reflection, when men try to find a place for it in their cosmic economy; for its essence is poetical, expressive, contemplative, and it grows more and more nonsensical the more people insist on making it a prosaic, commercial exchange of views between two interlocutors.

Prayer, then, is private in two of its uses: as a call for help or a plea for something desired, and as an overflow of gratitude or a personal expression of emotion. But prayer in its public, ritual use, like religion as a whole, is *communal.* And this is basic to the distinction between prayer and magic, for magic is *individual.* The magician weaves his spells in order to gain personal ends; his chief concern with others is to use them as means for his own gratifications. In refutation of this description of magic it might be pointed out that rites like the Black Mass, which is the mass performed backwards before an inverted cross, are sometimes thought of as magical, and those who practice such rites regard themselves as magicians or witches. Such practices are, of course, communal, not individual. But they are not magic, even though the celebrants are magicians: they are the ceremonies of devil worship, of the veneration of Satan. Witchcraft in the Christian era has often been connected with these rites, for Satan is supposed to give witches their power. A witch practices her *craft* for individual ends, but she gathers with others of her kind at ceremonies like the Witches' Sabbath to practice her *religion.* Implicitly, the devil worshiper says, "Satan, be Thou my God!" So devil worship may be thought of as a religion; and it is necessarily heretical in that Satan is defined with reference to God, and the very existence of Satanism presupposes another religion of the larger community in which God is worshiped.

Another difference between magic and religion is that magic may be, and often is, directed against a whole society or any of its members, so that the magician attains his ends at the expense of others. Religion is concerned with the welfare of all, or at least with the welfare of all its communicants. This does not mean, in practice, that religion always supports all other social institutions. We have noted that even among primitives it is sometimes at odds with one or another institution, and the history of the modern world shows that religion may be revolutionary. The ethic of Calvinism, for instance, helped develop the young capitalist economy which opposed mercantilism, and Cromwell and his Puritan forces overthrew the English monarchy. But even the support of a revolution by religion is in the name of the whole community.

Many religions have elements of magic in them, but these are usually so thoroughly assimilated into the body of the religion that they no longer have the special characteristics that magic has when practiced outside of religion. So the distinctions we have just made hold essentially between religion, with whatever of magic it contains, and magic as it exists by itself. Typical of magical elements in religion is what is called *shamanism*. This is a generic name for religions in which the spirits or gods are conceived to be responsive only to those who have the power to make personal contact with the supernatural world, as spiritualistic mediums in our society profess to do. Shamans are a combination of priest and magician. They perform incantations, spells, and rituals which force the spirits to do their bidding, and in this they are magicians. But their behavior is prescribed by their religion and they use their power for the welfare of the people; in this they are priests. In shamanism, magic is devoid of individual, antisocial ends. It differs from prayer only in commanding instead of pleading.

Magic in itself, not as an element in religion, is in some ways more like science than it is like religion. In the Middle Ages and the early Renaissance many things that are now called magic (astrology and alchemy, for example) were attempts to create sciences.[17] Some of the greatest scientists of the early modern world, like Tycho Brahe and Kepler, were astrologers as well as astronomers. After all, science in its applications is, like magic, a road to power. And in the popular mind of the Renaissance, science was thought of as a kind of magic.[18]

[17] For an admirable exposition of the whole problem of magic and science, see two studies by Lynn Thorndike: *The Place of Magic in the Intellectual History of Europe*, Columbia Univ. Press, 1905, and *A History of Magic and Experimental Science*, Columbia Univ. Press, 1923-1941.
[18] See pp. 97-98.

One of the strangest of men, Paracelsus,[19] was both scientist and sorcerer in the practice of medicine. He opposed the theory that disease comes from the "humors" of the body. He introduced arsenic, opium, sulphur, and mercury as medicines. He made the first great study of the diseases of women. Yet he was avowedly a magician. As an alchemist he sought the Elixir, a single nostrum to cure all ills. In the course of his work, however, he decided that there was no one remedy for everything but that each disease could be treated by a specific medicine. And he initiated medical research in chemistry as the way to find such cures. In Paracelsus the practice of magic starts to become the practice of science.

In both magic and science men try to attain their ends by manipulation, by controlling forces external to themselves. In contrast, prayer is an attempt to gain those ends by supplication. In terms of their *attitude* to means, then, magic and science are like each other and unlike prayer; in terms of ends, all three may be alike. But in respect to the *nature* of the means, magic and religion are like each other and unlike science, for magic and religion deal with the supernatural while science deals with only the natural.

A final distinction of importance between magic and religion is that religion is both a means and an end, whereas magic is only a means. Religion is practiced not only because it is supposed to yield a variety of ends (such as salvation, virtue, power, and love), but because it is thought to be good and true in itself. No such consideration holds for magic, and there is no moral imperative to practice it. There is merely the magician's desire to attain ends which are beyond his normal powers or are unsanctioned by his society.

The relation, then, between magic and religion is best thought of as a curve or line with magic at one end and religion at the other and differing combinations of the two in between. But in their pure forms religion and magic are quite different in emphasis. These differences may be stated briefly in summary: magic is individual and religion communal; magic may be antisocial but religion supports society; magic manipulates and religion supplicates; and religion is an end as well as a means but magic is only a means. There are further differences, to be sure. Religious ritual, for example, is performed at

[19] Whose life was probably the basis for the Faust legend, which so intrigued Goethe and Marlowe.

fixed times while magic is practiced whenever the magician chooses. But such differences are implied by what has already been said.

The history of civilization exhibits a constant secularization of the ideas and doctrines of religion. What was at one time sacred, holy, and extraordinary becomes at a later time secular, material, and ordinary. What was religious to Homer and Aeschylus, Augustine and Aquinas, Luther and Calvin is often a secular problem for contemporary philosophers and social scientists. The religious argument that all men are brothers because they are children of the same Heavenly Father reappears in secular guise as a belief in equality because blood does not vary from race to race, genetic abilities are equally distributed through the world, and variations in intelligence and ability depend on culture, not on race, religion, or nationality.

When religious ideas are secularized the special emotional quality of religion which was connected with them usually disappears. Sometimes, however, the opposite takes place: the secular becomes the religious and takes on a stronger emotional tone. So when sensitive and intelligent people who need the certainty and emotional security of religion find they cannot accept the beliefs of any church, they may make something secular into a religion. In the nineteenth century, for example, there was a Religion of Science and a Religion of Art. It is probable that each of these satisfied somewhat different religious needs. The Religion of Science attracted those who needed religion essentially as faith or belief. The Religion of Art attracted those who needed religion essentially as emotional commitment and moral guidance.

Unfortunately for their status as religion, and fortunately for their own integrity and development, neither science nor art can meet the demands that are made of religion. Science cannot give the certainty that religion offers, and it cannot answer in any way many questions raised by theology, questions of purpose and of final ends, for example. When science is treated as a religion its qualities as science are destroyed. Its conclusions are treated as final, which puts an end to the process of inquiry. Opposition becomes heresy. The errors of the moment (and there are many scientific errors at any moment) are enshrined as doctrine. And science loses its basically self-corrective character.

When religious demands are made of art the confusion of categories is quite as serious, although less obvious. There is a similarity between moments of aesthetic concentration and religious exaltation that leads people to identify the two. And there are often different levels of meaning in art as in religion, the symbols of both rarely standing for one thing unequivocally. In addition, art often relies heavily on symbols which still have religious associations. But those who make a religion of art are not satisfied with analogies. They want more: *the* truth, the source of morality, ultimate wisdom about man and his place in the cosmos. This is asking what art cannot give. To be sure, art contains insights into all these things, but the insights are those of the individual artist, and a higher authority cannot legitimately be claimed for them. Art is corrupted, as science is, when treated as religion. The essential individuality of every work of art is minimized and the artist is treated as an oracle, not as a man.

In the twentieth century, formal religious faith has declined (except for a limited mid-century revival in some countries), but there has been a corresponding lack of belief in science as an answer to all man's questions, and an increasing isolation of the artist from the community which is his potential audience. Yet ours is in a way one of the great ages of faith. Religious emotions and attitudes are now focused on political and social theories which promise some kind of redemption in this world. The two most important movements which are at once religious and political are fascism and communism. Both have promised man a virtual paradise on earth, both have been objects of fanatical devotion, and both have created ritual from public spectacle and prophet and priest from the political leader.

Fascism and communism are deeply hostile to formal religion—indeed, to anything that would divide the allegiance of their people. This suggests that they are to be regarded as surrogates for religion, demanding even more devotion than the Church inspired in the Middle Ages, for then the state was a permissible rival. "Render unto Caesar those things that are Caesar's, and unto God those things that are God's" allowed a duality of allegiance, although one in which God's claims were greater. The totalitarian dispensation asserts: "Render unto Caesar all things, for all are Caesar's."

In the content of these political religions many religious themes are restated in presumably secular terms. The Marxist philosophy of history, for example, echoes the Augustinian, which is the traditional Christian philosophy of history. For Augustine, man originally lives in Eden, the Earthly Paradise; for Marxism, man is originally in a state of primitive communism. In both, the goodness of man's life at the time is based on innocence, an unawareness of the possibilities of evil.[20] For Augustine, the Fall is brought about by the apple and the serpent, by man's disobedience to God's command and a consequent awareness of good and evil; for Marxism, the Fall results from surplus value which is created by technological advance, and by the resultant competition for surplus. For Augustine, history is essentially a struggle between good and evil; for Marxism, "the law of history is the law of the class struggle." For Augustine, the crucifixion of Christ and the emergence of the Church are decisive in the struggle and good then must surely triumph; for Marxism, the same end is brought about by the emergence of the proletariat and the formation of the Party. For Augustine, the conflict between good and evil is resolved on the greatest of battlefields, that of Armageddon; for Marxism, it is resolved in the Social Revolution. For Augustine, Armageddon is followed by the millennium which ends with the Last Judgment; for Marxism, the Social Revolution is followed by the dictatorship of the proletariat which ends in the classless society. And for Augustine this development is an inevitable sequence dictated by God. For Marxism it is an inevitable sequence dictated by the laws of history.

When a religious idea is secularized (and this is true in science and philosophy as well as in art and ethics) it functions in two contexts, its original religious context, in which it remains, and the new context which has adopted it. But when a political idea is treated as if it were religious or a religious idea is politicalized, and formal religious institutions are suppressed or persecuted, the idea exists in only one context, for it has been wrested out of the old to be placed in the new.

A secularized religious idea may be tested by scientific procedure, be analyzed anew, or gain an additional dimension of meaning because

[20] For Marx's collaborator, Friedrich Engels, men and women "before the division into classes" were "free and equal," characterized by "personal dignity, uprightness, strength of character, and courage." As for their physical qualities, many could even travel "farther and faster in twenty-four hours than a horse." Shades of Jean Jacques Rousseau!

of its new context.[21] So it is not necessarily worsened by being placed in a secular setting. But an initially secular idea which becomes religious loses many of the qualities which originally made it valuable. And when an initially religious idea—or any other—is politicalized and given religious force within the new context, it is an aid to despotism.

The enormous political advantage of supernational theology, that one can appeal from the injustice of the state to the moral authority of religion, is lost when the theory on which the state rests is treated as theology and the churches are deprived by law of even moral authority. The state is then a corrupt or perverted theocracy, despotic as all theocracies are,[22] but dependent on the caprice of its rulers, not on the unchanging word of God—even as interpreted by His presumed representatives.

These new theocratic societies boast of their unity, their cohesiveness, but in reality the social ties of religion have been replaced by the shackles of totalitarianism—and it is difficult to know whether the new dispensation gives society the same cohesiveness as the old because of the amount of force used to stifle dissent and to destroy dissenters. In any event, totalitarianism, when it eliminates or emasculates all churches, cannot be cited as evidence that religion is not a necessary institution: for all the emotional attributes of religion are attached to the body politic. That religion has been, in the past, a part of every civilization, and that it has been minimized in our day only where there is a religious surrogate, suggests a hypothesis. Perhaps it is not institutional religion that is necessary for civilization, but something else, of which institutional religion is the chief example in history.

It may be that what is indispensable is a type or *quality* of belief that must be shared by its members if a culture is to continue. Religion is not, we have contended, necessarily a set of theological or any other beliefs, but rather a way in which people act, or a way in which any set of beliefs is held. In so far as a quality of *belief* is singled out as necessary for cohesiveness in our day, a new sense of

[21] For example, the idea of progress. As expressed by eighteenth-century French *philosophes,* "the doctrine of progress," wrote Carl Becker, "was but a modification, however important, of the Christian doctrine of redemption."
[22] Theocracy (*theos,* God + *kratein,* rule by) is never rule by God, who does not come down to sit in the executive chair. It is rule by a church, in fact by a leader or leaders of that church, who proclaim themselves God's vicars on earth and will brook no opposition.

the word "myth" has been introduced. A myth, by this definition, is any belief, regardless of its content or its truth, which people accept as sufficiently sacred to dedicate themselves to it; it can give a feeling of participating in something more important than themselves. As a consequence, acceptance of a myth makes for social cohesiveness. This is the essentially *social* value of religion with which we have been concerned. And it is becoming increasingly clear that the calculation of individual advantage, even when "enlightened" (that is, when there is the realization that others must be allowed to seek their advantage if there is to be a system that guarantees one's own search for advantage), is not sufficient for the continuance of society.

Democracy and science would qualify as myths if they were held with sufficient devotion. But they are not. At this time they evoke in only a very few people the kind and quality of emotional commitment that religion does in many. And there are no other religious surrogates open to us now. So we probably could not today have the kind of cohesiveness in a democratic society without religion that one finds in societies where religion is rooted firmly. But it is possible to *conceive* of a society without institutional religion as we know it, in which cohesiveness is maintained by a widespread and deep faith in democracy and science, or still other values.

The twentieth century has suffered, in the Soviet Union, in Nazi Germany, and in their imitators, from the results of making the state the basic myth of a society. It may be that any myth yields cohesiveness, but we pay a terrible price in other values if we accept a myth only because it yields this one value. The *content* of the myth is important to the realization of many ideals. The myth of the state destroys freedom, individuality, the autonomy of art and science, and so on.

There has been a long tradition of deification of the state in the person of a king or emperor; and the unlettered readily accept the myth of the state. Would any other myth be better? Much might be gained if democracy were a central myth, but much might be lost. Individual religious values, which we have not discussed (for example, piety and humility), might vanish. But there might be a growth in man's tolerance of man and in his ability, intellectually, to control his environment and his destiny. Still, there is no evidence in history that democracy, or science for that matter, could ever be accepted as a myth by enough people in any culture for it to be focal and pervasive. And democracy and science have been embraced with fervor only among the best educated. Yet in their current forms they are relatively new to human history, and it may be, for good or ill, that a day will come when a sheerly moral, or procedural,[23] or scientific myth will engage the affections and emotions of people sufficiently to keep their culture essentially intact and continuous. On the basis of what we know now, however, we cannot predict that there will ever be such a time, and if there is, whether the advantages will outweigh the disadvantages.

When we discuss ritual we must be wary of assuming too much uniformity within any religion. Ritual contains no intellectual issues and so does not lead to controversy about ideas, as theology—and even myth—does. But there is sometimes great dispute about whether—and how—any particular ritual should be performed, or whether ritual should exist at all. This may be overlooked if one concentrates on primitive religion, because, lacking historic documents, primitive cultures appear totally homogeneous and unchanging. Still, the ancient religions of the West and the Near East have documents aplenty to make it clear that great struggles over ritual have existed.

"The letter killeth, but the spirit giveth life" is a cry raised time and again by those who find that ritual observance may be stultifying to the life of spirit. The issue arises when religion has become more than ritual, or when reflection on ritual has imbued it with spiritual and ethical significance. The chief struggle is between those for whom ritual is deeply important in itself or for its presumed efficacy, and those for whom it is significant only as an expression of the spiritual and moral,[24] for whom it is essentially symbolic, not magical. The leaders of the contending parties are priests on the one side and prophets on the other.

Religions are founded by prophets but governed by priests. (Mohammed, for example, is called the Prophet by his followers.) Not all prophets found religions, but all are innovators and potential revolutionaries, even when the innovations are based on their ideas of a past whose theory or practice is neglected. Thus Luther and Calvin went back to the Scriptures and to Saint Augustine, eventually creat-

[23] Such as intellectual method, or reason.

[24] It is also implicitly between supporters of the communal, in which men are differentiated only by their status, and supporters of the individual, in which even community depends on differences in persons.

ing a rupture with the Roman Catholic Church and establishing new religions. John Wesley, who founded the Methodist Episcopal Church, had the same intellectual antecedents, and was originally a reformer who wanted the Anglican Church to return to its sources. All three were secondary, or derivative, prophets. Their arguments depended on the records of earlier, primary prophets who claimed to speak the word of God as a result of direct revelation.

A prophet [25] does one of two things. He denounces his correligionists for lack of adherence or mere mechanical adherence to the ritual, urging meaningful participation in it; or he enjoins them to regard the ritual less and the spirit more. No society is ever spiritual enough for a Jeremiah or an Isaiah, although it may be ritualistic enough. When a society is ritualistic the prophet will probably urge neglect of ritual for the sake of spirit; when a society is not ritualistic he may urge a return to ritual for the sake of spirit.

The Old Testament is by far the most complete source of information about the opposition between priest and prophet. At first priest and prophet are identical, or closely connected, because the prophet who creates a new religion or leads it in times of trouble is the natural choice for the position of priest in that religion. Moses and Samuel, for example, were both priests and prophets. But often the prophet is not a priest. And even if he is, he may be at variance with the priests as a class, and suffer exile or death. The distinction between priest and prophet was kept alive by the tradition of separate prophetic and priestly oracles. The latter were a method of divination which could be taught and passed on from generation to generation, the former an individual and perhaps mystical matter. Later there was even a class, or guild, of soothsayers who practiced a kind of magic and were opposed by both prophets and priests.

The essence of prophetic religion for the Jewish prophets was to know and obey God. Traditionally, from the time of Ezra, God's command included the performance of ritual; therefore ritual was not merely observance for its own sake, but was also a fulfillment of God's will. So to Malachi it was important that Israel return to the ritual sacrifice it had neglected. Yet the language of Isaiah and Amos, prophets of an earlier and more ritualistic period, is to the opposite effect. "To what purpose is the multitude of your sacrifices unto me?

[25] Except an imagined *first* prophet who preaches to a people with no religion of any kind.

saith the Lord. . . . Bring no more vain oblations. . . . Your new moons and your appointed feasts my soul hateth: they are a trouble unto me; I am weary to bear them."[26] "I hate, I despise your feast days. . . . Take thou away from me the noise of thy songs; for I will not hear the melody of thy viols. But let judgment run down as waters, and righteousness as a mighty stream."[27]

A prophet may be successful in several ways: his injunctions may become reforms and he himself may enter the priestly class, perhaps as its leader. Then it is likely that new ritual practice will be based on what he has said and done, and after a time will become so fully institutionalized that the spirit is again lost and another prophet arises. There is a direct analogy of this religious sequence to a sequence in political behavior. A successful revolution becomes institutionalized as a new authority which in turn breeds further rebellion. But the success of a prophet may be of a different kind and have other consequences: he may establish a sort of prophetic dispensation running parallel to the priestly dispensation but neither absorbing nor being absorbed by it. This, in fact, was what happened with most of the Old Testament prophets. In their case it was the more likely of the two possibilities[28] because they understood their mission as that of bringing man directly to God; in a way they substituted their own mediation as God's spokesmen for the traditional mediation of church and ritual, and perhaps envisioned a time when all men would become prophets, when church and ritual would be unnecessary. So they did not want, like most political revolutionaries, to seize power, but rather to coexist with formal authority, and perhaps ultimately to transcend it. If there is a political analogy to these Hebrew prophets, it is that of anarchists who hope to eliminate political power altogether, not merely to take it for themselves.

Prophecy is usually thought of today as prediction of inescapable future events. It is not ordinary prediction, which might be scientific, but it is presumed knowledge of things that *must* happen. Scientific prediction is based on evidence and makes no pretense of certainty. Prophecy of the future is based either on an essentially magical method of auguries, about which the predictor admits no argument, or on revelation. Old Testament prophets, however, usually made

[26] Isaiah 1:11, 13, 14.
[27] Amos 5:21, 23, 24.
[28] It is not always the more likely.

predictions about what would occur as a consequence of particular human behavior, especially moral behavior, so the predicted events were not inevitable. Fatalism, a doctrine held implicitly by many soothsayers and magicians, is the belief, as we have said, that certain events will occur no matter what has preceded them. Determinism—and the prophets were moral determinists—is the belief that certain events will occur *if and only if* they are preceded by certain other events or behavior.[29] It follows that if men heed the prophet's warning and behave as they ought, they will create a different destiny.

Prophecy in the major religions of the West is not only prediction but also the utterance of God's commands. In this especially, the Old Testament prophet differed from the soothsayer and the priest, because the soothsayer was concerned only with the future event, and the priest interpreted existing doctrine. The prophet stated the will of God as directly revealed to him and called for a change in present behavior. Obviously, then, there is some mystical element in prophecy, and the prophet—or rather the primary prophet—is a special type of mystic. The mystic has personal contact with the supernatural in which he may only experience a supernatural presence, or may gain knowledge from it. In either case the mystical experience is for the mystic alone unless he is also a prophet, in which case he receives a message to be passed on to others. Prophets, therefore, have a claim to leadership which other mystics usually do not have. "Moreover the word of the Lord came to me, saying, Go and cry in the ears of Jerusalem, saying, Thus saith the Lord." [30]

But the mystic, if not so troublesome to the hierarchy of Western religions as the prophet, is disturbing enough. Like the primary prophet, he claims direct access to the source of authority, God, and he cannot be controverted by rational argument or appeals to tradition, history, or worldly authority. The priest does not know whether a man who claims to be a mystic is, by the priest's own lights, telling the truth. If he is not, he is probably a scoundrel and may preach heresy when it is to his advantage. If he is, he does not need the mediation of the rituals over which the priest presides, and he may at any time become a prophet, preaching God's word.

Religion in the modern world differs from medieval religion

[29] For a fuller discussion of fatalism and determinism, the reader is referred to Chap. 8, pp. 149-54, on "determinism, fatalism, and freedom."
[30] Jeremiah 2:1, 2.

chiefly because of two events: the Reformation, which brought the Protestant churches into being, and the Catholic Reformation, or Counter Reformation, which eliminated many abuses and tightened Roman Catholic dogma. Both events led to puritanism, which most contemporaries take for granted as part of religion. After all, it is an exaltation of the spirit and a denigration of the flesh. And what could seem more typical of religion than an emphasis on the spirit, not the flesh?

It is usually forgotten that denigration of the flesh characterized many of the great medieval heresies. The Cathar and the Albigensian heresies, in particular, had a strong admixture of Manicheism, which held that Satan is a Power equal, or almost equal, to God. The Cathari and the Albigenses wanted to be purer than the pure; they were enthusiasts in the original sense. Purity, they thought, consisted in living as much as possible for the spirit alone. The kingdom of matter, which included the flesh, was created and ruled by Satan. The kingdom of the spirit was God's. So they objected even to rites like the Eucharist, in which God was supposed to be present in the Host. For the Host was material and could not, even after consecration, be God or His symbol.

Cathari and Albigenses at the highest stage of renunciation of the flesh were called "the perfect." And all were expected to strive for that consummation. Yet along the way many strange practices were permitted, perhaps even encouraged. Sexuality was a sin but if one could not resist he could at least be homosexual, which was not so bad as yielding to the great temptation, women. Or where homosexuality was regarded as even more degrading than heterosexuality, it probably was thought to express a proper disdain for the flesh.

Not all "enthusiasm" has such unpleasant consequences, of course, but when it is outside the norms of conduct for a society, it more readily produces them. After all, when a type of conduct conforms to the mores of a culture it is not likely to lead to degradation of the personality, no matter how bad that conduct is from the standpoint of another culture. A cannibal in a cannibal tribe is like all his fellows. There is no consequence of the cannibalism on the rest of his behavior. But a cannibal in our society would be such a monster of iniquity—and know he was—that he would probably be vicious in his other practices.

In Western civilization mystics are rare, and it is normally believed that the rareness of their experience makes it impossible to communicate anything about it except to other mystics. Even when two mystics are friends, like St. Theresa and St. John of the Cross, each of them has his vision of God alone. Thus they stand apart from the rest of the religious community, for whom observances and common worship are necessary. Yet there are other societies in which standard religious experience is essentially mystical, sometimes in the central rite. Of course, in order for mystical experiences to be part of religious ritual many people must have them; and in these religions many people do.

The Crow Indians of Eastern Montana provide an interesting example of a religion oriented to mysticism.[31] It is believed among the Crow that success in life results from having a vision. Lowie quotes One-Blue-Bead: "When I was a boy I was poor. I saw war parties come back with leaders in front and having a procession. I used to envy them and made up my mind to fast and become like them. When I saw the vision I got what I had longed for . . . I killed eight enemies." Other Crows reaped different rewards, but these were almost always what they had wanted before the vision: wealth, prestige, a good-tempered wife. One-Blue-Bead assumed that fasting helped bring about the vision, and indeed it is one of the simpler methods of the Crow. Others range from mental concentration to torture and drugs.[32] The Crow, in order to have a vision, goes to a remote spot, usually near the top of a hill or mountain. He wears nothing but a breechclout during the day and covers himself with only a buffalo robe at night. He neither eats nor drinks and, as a kind of climax, tortures himself, commonly offering a finger-joint of the left hand as sacrifice. If successful (and the Crow seem rarely to lie about this), the Indian returns to the tribe with a power in him which everyone is quick to recognize.

Visions, trances, and mystical experiences are desired by most of the Indians of North America, except the Pueblos. One of the most important means of inducing these states is the eating of a narcotic called peyotl, a cactus button grown originally in Mexico, which gives sensations of levitation and brilliant color and has a powerful emo-

[31] As reported by Robert H. Lowie, *Primitive Religion*, Liveright, 1948.
[32] These methods are widespread and have been incorporated in the ritual of many Indian tribes.

tional effect, apparently never erotic.[33] The use of peyotl has become more and more ceremonial until now it is taken like the Host at the Mass and is identified with the god himself. Unlike the Mass, however, the peyotl ceremony brings the communicant a direct experience of the supernatural in the form of vision or hallucination.

Two important points can be made about this sort of religion. First, in solitary ritual like that of the Crow visionary, the communal aspect of religion is necessarily reduced, or varied. Yet the ritual method of inducing the vision is in a way social: it is prescribed and handed on from one generation to another. And the effect sought is in part social: it is realized on the return of the individual to the tribe. Thus the Crow regards his vision as only instrumental. In this he differs from the mystic of Western civilization who regards the mystical experience as an end in itself and finds the rest of life pale in comparison. The prophet differs from both. Like the Crow, he treats his vision or revelation as instrumental (although he may, in addition, cherish it as an end), for it gives him his message. But unlike the Crow or the mystic, he treats it as instrumental to the welfare of the whole group to whom he divulges the message.

The tiny number of mystics in Western civilization provide an individualistic leaven in communal worship. But when mysticism is the rule rather than the exception it takes its origin from, and plucks its fruit in, organized religion and society. And in the cult of the peyotl the individual visions of the communicants occur at the same time and in each other's presence. Hence, although the experience itself is not identical for all the celebrants (the visions differ somewhat), individual variation takes place on a common occasion.

The second point is that extreme forms of behavior are required if so highly individualized an experience as the mystical vision becomes an ordinary religious or social ideal. It is a rare person who is capable of mystical experience in the normal course of life. At those times in the medieval world when visions became a social desideratum, even abstinence, fasting, and prayer were insufficient aids for some, who resorted to the hair shirt and the lash. Because the Indians of the American plains expect an even larger proportion of the population to have visions, they have had to institutionalize alcohol, drugs, and

[33] Aldous Huxley has described his responses to peyotl in *The Doors of Perception,* Harper, 1954.

torture as means of inducing them. And how can we trust visions brought about by such methods? Even if we grant that some visions are experiences of an external Reality, can we credit those that result from an abnormal bodily condition? "From a scientific point of view," said Bertrand Russell, "we can make no distinction between the man who eats little and sees heaven and the man who drinks much and sees snakes."

The vision of the lonely Crow on the mountain is a way in which a source of power is tapped. Part of our definition of religion was recognition of power beyond our ordinary selves. So the behavior of the Crow, which seems so different from our own religious practices, is quite similar in underlying intent. The proper performance of ritual —or God's grace—gives us a strength greater than before. We are enabled to carry through undertakings in which we would otherwise falter. An interesting parallel is found in the early religion of Rome, to which we have referred. Even the simple religion of the family had a ritual that was fixed, and any flaw or mistake in its performance was supposed to prevent the numen, or power beyond themselves, from entering their lives.

The Indian visionary may be regarded, therefore, as participating in a rite of a special kind rather than as departing from rite altogether in his ordinary religious observance. Perhaps the chief difference between the two types of ritual is that precise observance in a more fully communal religion is expected to be successful in making contact with the deity, whereas the Crow assumes that many will fail in the attempt to attain a vision, which is his way of making contact. The difficulties of institutionalizing so individual and aberrant a thing as the vision reflect certain weaknesses in Crow and other Indian societies. It is as if every man were expected to be a hero, so that failure to wear the Congressional Medal of Honor at a dinner party might cause one never to be invited again.

Actually, the type of hero the Indian visionary represents is found in the legends of many cultures. He is the man who surmounts numerous difficulties and undergoes great suffering to attain some treasure, usually from a remote place, which he brings back to his society. In the fairy tales of many nations a number of princes seek fabulous objects which they take to the princess they are all courting, and the prince who returns with the most marvelous one marries the princess and inherits the kingdom. In more austere legend it is wisdom, or purity, or even mystical experience, that the hero seeks in almost inaccessible places. But the hero is usually, by definition, an exceptional man. To try to make a society of heroes almost necessarily excludes some men from a common dispensation.

We have implied that much of what was said about religion also held true for society. And there is at least an analogy: society has its men of original views as religion has its mystics. Those views may bind men more tightly to social tradition. They can, for example, be insights into the meaning and value of convention and tradition. Or they can set them apart from their neighbors in matters of belief. The man who is set apart may, like the mystic, want simply to be left alone to follow his own lights or he may, like the prophet, feel that his ideas should be accepted by others. His opponents will be the "priests" of his society, men who demand social observance which they either interpret in the old way or defend without reflection or explanation. And the conflict between the two will fall into the same pattern as the conflict between priest and prophet.

There are societies which have developed a theory of individualism in some ways like the theory of mysticism among the Crow. Every man is expected to be quite different from his neighbors in attitude and idea. And anyone whose originality is great enough is expected to reap the social rewards of power, wealth, and prestige. Yet in order that the society be held together at all there has to be a kind of institutionalization of individuality, a situation difficult to define or attain. The United States has a tradition of this kind, and it exists side by side with a tradition of conformity to public opinion and the habits of the majority. The conflict between the two develops tensions and paradoxes evident to the discerning visitor, like Tocqueville, but hidden from most of us, who share them.

# 11

# Art and Its Social Functions

In order to communicate about what they perceive, men use symbols to name or otherwise designate identifiable and recurrent bits of the flux of experience. Because we use symbols to stand for things, we can think about those things even when they are not present to our senses, even when they have never been present to our senses or anyone's senses. Mermaids do not have to exist in order to be talked about intelligibly, but the word which stands for them has to exist.

The basic function performed by symbols which stand for the objects of our experience is to identify and define them for ourselves and others. To some extent, this process of identification standardizes perception. When we have words for something, we know what we can expect to see when we look at it. And, within limits, we do see what we expect to see. Countless psychological experiments on the completion of common sequences show that even when the entire sequence is not present we tend to see it as though it were. If we are presented very briefly with the letters ELIZABE**, many of us tend to see the entire name ELIZABETH. If the sequence is an even more common one, ENGLAND'S QUEEN ELIZABE**, still more people see it as if it were completed. The same tendency to complete sequences or patterns is found when we are confronted for only an instant with pictures of houses, animals, human faces, in which one or more details is missing.

Usually, then, when we call attention to an object by using its name, it will be seen by others as it is seen by us, even when it does not correspond exactly to the name by which we call it. The use of symbols helps to standardize perception, but in the process it often falsifies perception. Most perception is stereotyped. It fits preconceived notions, which are created for the most part by symbols. One of the talents and functions of the artist, we shall argue later, is to destroy the stereotype by perceiving freshly and individually, with a minimum of preconceptions.

Symbols not only name the objects of experience, but they also place them in a context. And the context may alter response to the object, the evaluation of it, even the way it is perceived. In an experiment on responses to the expressions on faces, the same photograph of a grimacing face was presented to two groups of subjects. The first group was told that the photograph was taken while its subject was viewing a hanging, and it characterized the expression on the face as "disgust" or "anxiety." The second group was told that the photograph was snapped as its subject completed a hundred-yard dash. This group found the expression to be such things as "effort" or "determination." When similar photographs were shown people who were told nothing whatsoever about the context, they were either bewildered and incapable of characterizing the expression at all, or they described it in widely different manners. We are not in fact used to dealing intellectually with discrete events but rather with events in sequence or in context. And very often we actually perceive only discrete events, but know the sequence or context because we are told them. Experiments make it clear that perception of such events is itself altered by the sequence or context in which they are placed by symbols. Moreover, when several people have the same experience, their discussion of it may change their memory of what they saw. More interesting as a psychological phenomenon, if they have that experience again there is a tendency to see it differently from the way it was perceived the first time and more as they remembered it after talking about it. In fact, there is even a tendency for a norm to be established, so that several quite different individual perceptions of the same thing tend to converge during a second experience in which all the individuals are present at once.

Muzafer Sherif performed a significant experiment on the autokinetic phenomenon, which is produced by looking at a point of light in an otherwise totally dark room. Since there is no other object visible to which the point of light can be related in the visual field, it is seen to move, although it is in fact perfectly stationary. The distance of the perceived movement varies enormously with observers, from the slightest movement to a very considerable one. Sherif chose three men who had experienced the phenomenon independently and with a great difference in the amount of movement they saw. He then placed them in the room at the same time and had each one in turn describe aloud what he saw. The reports were very much more alike than they had been when given separately. Each of the three was then tested separately and continued to give the report he had made when the others were present, rather than his first independent report. Apparently, once group norms are established, they persist, even in isolation. And norms are not established only when people undergo an experience together. On the contrary, most norms are fixed in us in childhood as we are socialized. And socialization is, after all, a process of implicit and explicit instruction carried on for the most part through the medium of language. Some differences in expectation exist in different cultures and in different historical periods. So, since we tend to see what we expect to see, there must be some differences in perception from culture to culture and period to period.

We have commented that language embodies the values of a culture. Now we must add that those values, most powerful perhaps when they are only implicit, affect perception. The men of an African tribe that finds women beautiful when they are very fat and covered with grease probably do not, literally, see them as we would. It is not the case that they and we see exactly the same thing but appraise it differently. If they described a particular woman whom they found "beautiful" and then we described the same woman, the descriptive adjectives would probably be highly evaluative and would be evidence of a difference in perception. If they spoke English, they might talk of her skin which had been rubbed with rancid oils as "gleaming," "shining," "bright," or "glowing." The nicest adjective we would use might be "glistening," but we would more likely say her skin was "sweaty," shiny,"[1] or "wet." In each case the adjectives contain both description and evaluation, and they imply that we are attracted or repulsed. But the elements of description are not separable from those

[1] The great difference in the feeling given by the words "shining" and "shiny" is evident in many common uses. We compliment a lady by telling her that her eyes are shining (like stars), but we chide her when we say that her nose is shiny.

of evaluation. The adjectives characterize a woman on whose physical charms an evaluative judgment is being made; they are not intended to be an unemotional scientific description. If the latter were attempted, we might get quite similar reports of the woman's appearance from both Africans and Americans. But that would be a result of the kind of conscious effort—and perhaps of training—that is suitable to an observer in a laboratory. And it requires not only care and training in the use of words, but also care and training in observation, that is, in perception itself. So the tone of the original descriptions, probably like the perceptions on which they were based, contains our responses to and judgments of the object.

The symbols that carry social norms and values are not only words. The norms of perception are conveyed and reinforced by symbols in the visual arts—by posters, magazine covers, photographs, motion pictures, paintings, and statues. In addition, the values and perspectives of a society are implicit in all the arts of that society, and perspectives and values are closely related to perception. The vividness and emotional power of the arts make them important sources of social norms. Thus the arts are a great force in the creation of a society and a culture.

Social norms are, of course, indispensable to any society and are valuable even in giving direction to new thought and providing a base for new perspectives and values, which are most fruitful socially when they are related to older ones. But social norms also hold thought in well-grooved channels and sometimes substitute banal or falsified stereotypes for original thought and fresh perception. Creative work in mathematics, for example, usually requires startlingly new concepts and approaches. Most of it is done by men in their twenties who are not yet totally submerged in conventional thought.

Paradoxically, art [2] not only reinforces social norms but destroys them. Within indefinable cultural limits, art is highly individual, and its individuality is dual. First, it focuses on individual objects: this man, this colored surface, this sound. Second, the artist must see each object in an individual way, almost as if it had never been seen before. An artist may consciously relate himself to his forebears and deliberately perpetuate a tradition. This is no deterrent to novelty. It is the

[2] The word "art" is used throughout to mean the literary, visual, and musical arts.

use he makes of his tradition, the way in which he perpetuates it, that is his alone. No matter how he uses it, if he is not the merest imitator, his use constitutes a criticism of it. This is true even if he approves deeply of the tradition, for the very freshness of artistic perception presupposes and implies fresh evaluations and perspectives. And the vividness and power in art which make it so effective in reinforcing the social norms it contains make it equally effective in offering alternatives to the norms it opposes. The new perspective or perception that is most striking and rendered with the greatest artistry is most successful in creating a revolution in social thought and attitudes. Like all successful revolutions, these become the conservatism, even the academicism, of the next age.

Art has a special importance for the student of society. Social norms and values are for the most part implicit, and people who try to state them often do so incorrectly. The arts do not state explicitly; they preserve the very perceptions and the perspectives in which norms and values are implicit. Thus they constitute a perhaps unique record of any culture.

Now, symbols can be used for a great variety of purposes. They may, for example, make up a poem, a painting, or a song. But even in their furthest remove from purely cognitive statement, symbols contain implicit meanings that can be discovered. It is an old rule of logic that only propositions convey units of meaning, because only propositions can be true or false. So the command "Shut the window" is regarded as capable of evoking response but as, literally, conveying no information. Yet even on cursory analysis one realizes that two propositions are implicit in this command, presupposed by it if it is to yield response: "There is a window," and "The window is open." Equally, evocative or expressive statements in poetry or designs on a canvas may literally convey no meaning, and yet imply meanings. These meanings are not so obvious as those underlying "Shut the window," but they, too, are discoverable. As a simple analogy from the field of vision, if a table is described or painted exactly as it is seen, the description implies the *physical* point of view from which the table was looked at, for from any other point of view it would have been seen somewhat differently. The *intellectual* "point of view" implicit in a work of art is harder to get at. It is like the general hypothesis that underlies particular statements in science; but it too can be discovered with some degree of probability.

Art communicates, but it is more than communication. It conserves and reinforces social norms, but it destroys them. It is highly traditional and depends on the skills and techniques, the forms and patterns of the past, but it is good only when it is individual, fresh, and novel. It conveys meanings about all of human experience, but its worth is judged to a considerable extent by its internal structure. Finally, art is probably indispensable to society and yet is always a danger to the accepted, the established, and the static, so much so that it is always censored.[3] These are paradoxes that bedevil art criticism and the sociology of art. We will see how far we can go toward explaining them.

We have said that art is very much more than a form of communication. This is as true of the arts of language—poetry, drama, fiction—as it is of painting or music. Art yields an experience of the kind we call aesthetic, an experience most of us have in the presence of beauty, which gives deep satisfactions. Exactly why we have these satisfactions has puzzled philosophers for centuries, but it seems clear that they depend somehow on the qualities and organization of a work of art, including its meanings, not on the meanings in isolation. A good poem informing the reader that summer follows spring is not gratifying because of its information alone.

Art is particular and concrete, where science is universal and abstract. Yet the particular in art—the character in a novel or play, the object described in a poem or painted on a canvas—may also be a symbol of the universal. Unlike scientific symbols, the symbols of art are successful only when they are more than symbols, when they are valuable also in themselves, as particulars. In *Don Quixote,* Cervantes uses the sorrowful knight as a symbol of the unrealistic idealist in a materialist world. But if Quixote were only a symbol, no matter how successful a symbol, he would be an unsuccessful character and the novel a poor work of art. Quixote's fully realized individual character is a measure of Cervantes' stature as a novelist.

Works of art express, evoke, and are experiences. Their meanings are of the kind we find in all experience, and we *understand* art as we understand experience, not as we understand expository state-

[3] Censorship, implicit or explicit, always exists. When we oppose censorship it is not any censorship at all, though we sometimes talk as though that were meant. It is a particular degree of censorship that we oppose.

ment. Essentially, statement contains explicit and limited meaning or meanings, which, of course, logically imply other meanings; experience and art contain meaning implicitly, and art contains many meanings in addition to those consciously intended by the artist. Also, like other experience, art embodies different meanings, depending in part on the audience's background, point of view, and interests.

Although art yields an experience, it is an experience deliberately created by man, and it communicates. Its meanings, unlike those of nature, were put into it. They are meanings of various kinds and for purposes of aesthetic understanding they cannot be isolated intellectually from the work as a whole. Of course, literary art, which is made of language, usually conveys ideas and information; these have their full meaning in the way they function for the rest of the work. But they *can* be considered in their partial, expository meaning. When Keats writes, "I cannot tell what flowers are at my feet," he is clearly conveying information, in this case his lack of it. Taken apart from the rest of the poem—which it should not be if we want to understand its full meaning—this line would be simple exposition. And the author's ability to state his ideas in expository form when statement is required, as it is in Keats's line or in dialogue, is part of his ability as an artist, but only one part.

The nonliterary arts may convey information too. Portraits tell us what clothing the subject wore and what his features were like; cartoons can even tell a story. But this is very limited as information. And instrumental music conveys no information whatsoever. Yet when we view a painting or listen to music, we feel that an act of communication has taken place. If we are asked, "What was communicated?" we often cannot answer, and this may make us think that we were misled by our feelings. What is implied by the question is that any act of communication can be restated, or translated into linguistic communication. And this is not so. Listen to a piece of instrumental music that has no program notes attached to it and see if you can paraphrase what was communicated. Whatever you say in paraphrase will communicate very much less than, or even different things from, the music itself. But there is a clear sense in which the visual arts and music do communicate: they imply values and the perspectives from which things are seen.

The artist expresses value and perspective in many ways. Among these are composition and color in painting, theme and development in music, image and metaphor in poetry. When Shakespeare describes branches in autumn as "Bare, ruined choirs where late the sweet birds sang," he is comparing the leafless, birdless trees with empty choirs, and in this comparison he is telling us something about autumn trees that would take many lines of exposition to say, without, as we shall see, ever saying exactly the same thing. The relations of the images in a poem and the preponderance of one type of imagery rather than another also convey meaning. But that meaning, like all aesthetic meaning, is impossible to paraphrase fully and exactly because it is like the meaning of the man carrying the pail of water (see p. 158). It is embodied in something that is experienced, not just understood; and it is understood as part of the experienced whole, with all the color and vitality the artist has given that whole. When we paraphrase, we lose in meaning what the scientist did when he specified the weight of the pail. We substitute the purely symbolic meaning of science for the concreteness of art and experience, we direct attention to particular selected meanings as if there were no others, and we wipe out much that is implicit in the experience.

We ask of exposition, "What does this mean?" or, in the language of logic, "What is the proposition that this sentence conveys?" For logic, the proposition is the important concern and the sentence that bears it is irrelevant, except that it may be more or less lucid and convenient. But when we read or hear a poem, the line or sentence is not just a bearer of meaning: it is to be experienced. Hence any sentence that conveys a given proposition is, for the purposes of logic, as good as any other; but no word or line of a good poem can be changed without aesthetic loss. The meaning of a poem is not simply apprehended intellectually. It is *felt* as much as the other elements—sound, rhythm, and rhyme—and indeed if the poem had no meaning there would be little emotional response to it.

In an object of art, emotion is charged with meaning as meaning is with emotion. Felt meaning reveals and illuminates as it cannot in abstraction from its emotional context. How often, when we have "known" something for years, do we have an experience which reveals that knowledge to us for the first time in a living context, and suddenly we feel, "now I understand it"? The French actor-director Jean-Louis Barrault thought of "knowing" in this latter sense when he wrote, "Knowing something means having forgotten it and having found it again, inside. It is a 'digested' knowledge. By study we get to the core of a thing, we know it, then we forget it; at last, we find it again inside ourselves. From that moment on, we *know* the thing." Even sensuous qualities and textures have a new value and meaning in the context of a work of art. What is the sensuous value of a perfectly sung high C outside an aria in which it has a significant place? How different we find a shade of red on a canvas where it is contrasted with another red!

Since meaning in a work of art is inseparable from the work, it is a mistake to think that a poet could have embodied the meaning of one poem in a different poem had he chosen to do so. Meaning is not even expressed only, or completely, by words, but depends on the entire rhythmic effect and on individual sounds, and the rhythm and sounds are partially created by the meaning. Here is a poem by William Butler Yeats: [4]

> *Speech after long silence; it is right,*
> *All other lovers being estranged or dead,*
> *Unfriendly lamplight hid under its shade,*
> *The curtains drawn upon unfriendly night,*
> *That we descant and yet again descant*
> *Upon the supreme theme of Art and Song:*
> *Bodily decrepitude is wisdom; young*
> *We loved each other and were ignorant.*

The poem is made up of two quatrains which are not separated. Although the lines scan unevenly, each quatrain ends with a line of iambic pentameter, and the second quatrain opens with the same beat, which thus provides a regular metrical framework. Almost any literate person reading the opening line of the poem would place the first stress on the word "speech," give a light beat to both syllables of "after," and place the second stress on "long," the third on the first syllable of "silence." Thus: Spēēch ăftĕr lōng silēnce. The unstressed syllables of "after," followed by the stressed "long," show how very long the silence in fact was. The long vowel in the word "long" makes the silence even longer, as the stress on "speech" and the necessary pause before "after" make speech an abrupt change from silence. In a way, the rhythm answers the question: How long was this silence? To

[4] "After Long Silence" from *Collected Poems* by William Butler Yeats. Reprinted by permission of The Macmillan Company.

answer "twenty minutes" would not bear the intended meaning, which is a subjective feeling of time relative to a particular situation. So the sounds and rhythms bear part of the line's meaning. Conversely, had these four words been meaningless, nonsense words with the same placement of vowels and consonants, the line would have been nearly impossible to scan. Consider the nonsense line: "Freech aldem dong mipence," in which the words have the same length, the same vowels, and the diphthong sound of Yeats's line. An attempt to scan the nonsense line would most likely yield either three iambic feet or two anapestic, either Frĕech āldĕm dōng mĭpēnce, or Frĕech ăldēm dŏng mĭpēnce. Neither is correct for the real line. Rhythm, sound, and meaning are thus internally related, the rhythm and sound being determined to some extent by the meaning, and the words in a line so emphasized by the rhythm and sound as to clarify and enrich meaning.

But what about those portions of a work of art, especially literary art, which convey information? Characters in a novel or play make many statements which are either true or false. Well, the value of these is found in the way they reveal character, further the action, are related to the whole work. A character should not be confused with his creator. He reveals his creator's ideas by his functions in the structure of the work; he does not necessarily speak those ideas. The case is less obvious, but essentially the same, in a poem which contains no characters and in which the author seems to be addressing the reader directly. For it is not the author talking for himself; the "I" of the poem is, abstractly or dramatically, the Poet, not Mr. Wordsworth or Mr. Donne in their own persons.

Aristotle argued that a painter is not to be criticized for putting antlers on a doe. It is not the painter's task to copy on canvas the exact outlines of anything, nor the poet's task to utter true empirical propositions. They are expressing a particular view of the world as seen and evaluated from a particular perspective. If their purposes are served by deliberate distortion, that distortion is all to the good. The novelist Thomas Hardy wrote of a world in which fate seemed ultimately unkind, waiting for man to move himself painfully near his goal, and then destroying him. The world may not be such a place, but Hardy's novels are none the worse for that. They are to be judged in part by the way their own world is organized; by the consistency of spoken ideas with the character of their speaker; by the ways in which dialogue furthers the plot and is related to the over-all meanings of the book; by the relation of incident, character, and meaning. So it is not the truth of the ideas in themselves that is important but whether or not they are "true" *for* the characters and *to* the author's larger design.

A work of art expresses an organization of emotion or a part of the world seen from a perspective. Music may be thought of as expressing patterns of emotion, its variations, and development; musical development of a theme *may be* analogous to the progress of joy, sorrow, triumph, and so forth, through a series of emotional stages. Painting organizes forms and colors in emotional patterns, and, when it is representational, it views some part of the world in terms of principles and values of its own. When Impressionist painting first appeared in France, one of its characteristics was a concern with the effect of light on form, which included the way in which light is reflected from natural objects. The sharp outlines of natural shapes are dissolved under strong light and the light itself bursts from surfaces in patches of many colors. Sunlight on a leaf, viewed this way, shows not only lighter and darker greens but a variety of colors from yellow to purple.

Neoclassical painters like David and Ingres express the values of limitation and boundary, of precision and elegance—in the mathematical sense of excluding all elements that are not necessary for the basic structure—of the overcoming of tension by organization. In contrast, Romantic painters like Delacroix and Goya express the values of aspiration and excitement, of exuberance in the sense of richness beyond mere necessity, of the fascination of the grotesque and macabre—of the release of tension in conflict and violence.

Art produces and reproduces experience, but the experience it presents is not identical with any experience of the "subject" we may have apart from the work of art. The "subject" is filtered through the artist's sensitivity, and its elements are selected, emphasized, and organized in terms of the artist's own perspective on the world. The "content" of an oil painting of a still life is not the flowers or fruit on the table that the painter used as a model; the "content" of a poem is not the experience on which the poet based his lines. The *treatment* of the model or the original experience by the painter and the poet, and the *form* in which they place it, transform their own experience into art. In that transformation the subject of the original experience is metamorphosed by the requirements of the medium—oils, words—

the characteristics of the art form—a still life, a sonnet—and the artist's temperament, values, and perspective. The work of art has a "content" which is what the original "subject" has become after passing through the crucible of the artist's imagination. But this content is not separable from the form of the work, for it differs from the original subject precisely in being the result of, and incorporating, that form.

Sir Kenneth Clark, English art historian, wrote, " 'What is the nude?' It is an art form invented by the Greeks in the fifth century, just as opera is an art form invented in seventeenth-century Italy. The conclusion is certainly too abrupt, but it has the merit of emphasizing that the nude is not the subject of art, but a form of art." [5] The naked human body is the subject from which the concept of the nude was born. But comparison of the nude in art and the naked in the flesh quickly shows how different they are. And the enormous changes in the portrayal of the nude from one historical period to another do not mirror corresponding changes in the structure of the human body, but alterations in idea and ideal as well as alterations in artistic technique. The Greek female nude is idealized and usually heroic; she is a goddess. The medieval female nude is a mortal woman, often Eve.

Although art expresses the ideals and ideas of an age, each work of art is the product of a highly individuated human being; the stereotyped and the banal have no place in art. Craftsmanship alone does not make art, nor does sensitivity, for art is never just the expression of a culture or an age by a craftsman who grasps its essence. The artist must be a unique personality, able to express his uniqueness through his craft. The artist's individuality is expressed in the difference of his values, his emphases, his understanding, and his sensitivity to experience. This expression, to be sure, is unintelligible unless there is implied in his work some tradition which we can grasp. But in the great artist the traditional background and his individual differences from it are organized to make a distinct personality expressed in a personal perspective on the world. As we experience that world we are in contact with the personality it expresses; not only is there communication, but communion. The common locution, "the world of Shake-

[5] "The Naked and the Nude," *Art News,* October 1954, p. 20. Reprinted as chapter 1 of Kenneth Clark, *The Nude,* Doubleday, 1959. Perhaps "art form" is too strong a phrase for the nude, at least if we think of opera or of sculpture itself as an art form. The nude is a type of sculpture, painting, and graphic art, as Wagnerian music drama is a type of opera.

speare," or Dante, or Homer, is legitimate, for that world is our world seen and interpreted so individually and with such force and insight as to make it seem a world of its own. And we can enter it, seeing its detail, within the limits of our imagination and intelligence, very much as its creator did, for in his world we take on his values and perspective.

The artist's perspective is the meaning system in terms of which his art is to be interpreted. So we must know something of the values and conventional symbols of the artist's society as well as of the unique values and symbols of the artist. We often find that we must listen to a number of musical compositions, look at a number of canvases, or read a number of books by one man before we feel that we grasp the essential character of his work. To the novice, the works of Haydn and Mozart, or of Handel and Bach, sound alike because they have so many conventions of their own time in common. But after listening to a number of their works, it is easy to recognize the characteristic marks of each composer.

Is the artist's perspective, his meaning system, to be judged by the truth of what it conveys? No; even science, we have seen, is not judged entirely by its truth. Although the basic consideration about any scientific proposition is "Is it true?" there are other considerations: Does it lead to new ideas and experiments? Is it useful in organizing other concepts? Equivalents of these questions in art are: Is this part of a work useful in furthering the development of the work as a whole? Is this part consistent with the perspective of the rest of the work? And there are other questions, implicit in our meaning when we say such things as, "Chopin was a fine composer, but he didn't have the scope or stature of Bach." Most of us would agree that the world of Shakespeare has a range and profundity much greater than the world of John Keats, that Rembrandt had depth and insight far beyond Claude Monet. Some perspectives go further, reveal more, have greater richness than others.

Thus different perspectives in different works of art can be evaluated, but not in terms of scientific truth.[6] Science gives an intellectual order to our world so that we can find truth in it; art presents the variety of possible orderings of experience so we can find meaning

[6] The question whether art yields knowledge and, if so, what kind of knowledge, cannot be pursued here. But the interested reader may be referred to an article on the subject, "Art as Knowledge," by Ralph Ross, in *The Sewanee Review,* Autumn 1961.

and value in them. Art even offers syntheses of types of experience, and in the larger works, of all experience, yielding possible frameworks in which we can organize the fragmentary bits and pieces of our ordinary lives so that they are related and significant. Individual art forms may show the way life appears when viewed under the great categories of meaning. We see life as tragedy, life as comedy, life as character, life as aesthetic surface, and so on. The value of a particular artistic perspective, then, lies chiefly in its originality, depth, and incisiveness—perhaps, too, in its pertinence to our lives.

However individual an artist's work, it still shares with the work of other artists a fundamental style. It may be classical, romantic, baroque, impressionist, cubist, expressionist, but a fundamental style is that of an age, a civilization, a tradition. The artist's individuality is expressed in the way he uses a style, his combination of elements from different styles, his innovations. Styles are born, grow, develop, weaken, and die in the work of artists. So closely is each stage in a style connected to a time and place that historians of art, literature, and music date and place newly discovered works by (among other things) an examination of their style, and the diffusion of a style reveals the movements of peoples, trade, and war.

This connection of time, place, and artistic style is not accidental. The fundamental style of works of art, like the "life style" of a people, expresses the values of a culture, much of its world view, and its sense of itself. The aesthetic forms of rite and dance disappear when the ceremonies and festivities are forgotten. But aesthetic objects made by men often survive the civilization in which they were created. They afford us insights into those civilizations that could scarcely be acquired in any other way, for through their arts we learn how different civilizations conceived and perceived themselves and their world.

The development of a full artistic style is part of the maturing of a civilization, its discovery of itself, and its creation of a relatively unified culture. In its inception, it can borrow hugely from other civilizations, but it changes the meanings of what is borrowed, and slowly evolves its own forms. Christian art, for example, was originally part of an underground movement; it was an art of the catacombs. In its early history, Christian painting imitated Roman style and used the figures of antiquity, redefining them in its own terms, so that Hermes with the ram on his shoulder became the Good Shepherd carrying the lost sheep. It was not even important to early Christians whether a borrowed legend resembled the Christian story, as the tales of the Golden Age and Pandora in Hesiod resembled Eden and Eve. Any legend was useful that had interesting and well defined characters who could be transformed into something appropriately Christian, or who filled a gap in the Christian cosmogony.

Christian borrowing of detail from pagan antiquity continued for centuries. Dante and Milton used pagan figures to people a Christian universe. But the great medieval art styles made it unnecessary to express Christendom in visual pagan forms. One has only to glance at the Cathedral of Chartres to recognize the expression of a civilization utterly unlike those of Greece and Rome.

Art is as primary as religion and probably appears equally early in the development of human society. That men find some things beautiful and some ugly is as much a part of their nature as that they find some things sacred and some common. Human response to the beautiful is often very much like response to the sacred, and the two have been confused by many people. In the earliest religions the sacred and the aesthetic were fused rather than confused, and the fusion of the two was ritual.

The great experiences of life—birth, initiation, and marriage, for example—are made beautiful by art and sacramental by religion. Through this transformation a social function is performed: human experiences are celebrated, commemorated, and interpreted. The similarity in the social functions of art and religion is part of the explanation of the union of the two in ancient and primitive society. Also, those societies were almost coextensive with their religions, so that no aspect of human life was entirely secular. When art, therefore, celebrated or commemorated, or expressed the meaning of an aspect of human experience, it did so as part of religion, in which those functions were officially vested. In the course of history, as one after another of man's concerns were separated from religion—agriculture, government, education—art was separated from ritual and became the expression par excellence of the meaning of human experience and of the forms of celebration and commemoration. The secularization of art is part of the secularization of society. But in order to understand the social role of secular art it is valuable to examine the way art and religion were once fused in ritual.

Communal religious worship is a dramatic form of one sort or another, and it is fitting that drama itself came into being as an extension of ritual. Man's earliest tool was his own body, and he probably attained a mastery of its use in drama and dance before he developed any great skill in other arts. But as man learned better to ornament, paint, model, chant, and sing he used all these, not as embellishments of religious worship but as worship itself. Today we may see a clear distinction between the spiritual aspects of religion, like faith, religious contemplation, private prayer and communion, and aesthetic aspects of public ceremonial. The latter, not the former, is a fusion of religion and art, but the former often seems to us more essentially religious. That, however, is probably a view that developed comparatively late in human history. As society became more and more secular, religion became more and more spiritual—naturally enough, for it was turned away from the world.

Jane Ellen Harrison tells us, in her little book, *Ancient Art and Ritual,* that while we distinguish between a form of prayer and a work of art without any danger of confusion, the Huichol Indians find the two identical, for they paint their prayers. If they fear drought they paint a clay disk with a circle and arrows for the sun, yellow dots for cornfields, and curving lines for rain. When the disk is deposited on the altar of the godhouse, the prayer has been uttered.

When the ancient Greek wanted help from the Dioscuri, the "Saviors," he might carve a picture of them. Underneath he wrote the old word for prayer, *euche.* This was a sculptured prayer.

The need to pray may be satisfied by words spoken aloud or silently. It may also be satisfied by an action, like the rain dance, which is a musical and dramatic prayer. In very old religions, like the Egyptian worship of Osiris, prayers were acted and danced, painted and sculptured all at once. The mute meditation of the saint has its counterpart in the meditation and the prayer which emerges as a work of art and, as both prayer and art, is offered to the gods, to whom it was addressed.

Western art in our day is fundamentally secular. But before talking of it as such, it is worth noting how the creative experience of the artist has qualities very like those of religion, and how the aesthetic experience of the audience can lead to a religion of art. Let us consider one aspect of creativity first, and then move on to the audience.

A great deal of every artist's work is deliberate, and some artists are proportionately much more intellectual and deliberate than others (the enormous range can be seen in a contrast between, say, William Blake's poems and Matthew Arnold's); but, more or less, every artist is, in a way, led. There is a feeling that things come to him as though from a source outside his self, that they are caught, disciplined, and pressed into form.

The testimony of Socrates in the *Apology* is revealing. When he asked the poets the meaning of their works, they could answer less well than he. His conclusion that poets are the instruments of the gods, unaware of what they utter, overlooks the unfairness of the question. The poetic gift need not include a talent for exposition, nor need even the best of poets catch all his meanings in full consciousness. But there is a genuine insight in Socrates' explanation. And when, in the same dialogue, Socrates avows his own ignorance and attributes his wisdom to a daimon, a personal spirit, who whispered to him, he too is in a category with the poets. All creation, intellectual and artistic, has the same character in this respect: there is a leading, nearly a possession, in moments when the ideas come almost too fast to be captured in paint or ink or sound.

"Make me thy lyre," Shelley asks the West Wind, "even as the forest is." This is a request for renewed inspiration. Spirit originally meant breath; to be inspired was to be breathed into; and wind is a natural image for spirit. Shelley's line is not too far from that invocation of the Muse which was, in antiquity, more than conventional piety and is today more than pious convention. The spirit breathes into the poet, but the spirit need not be thought of as without; it may be deeply within. From what lies beneath consciousness something, based on memory, association, intuition, bursts into consciousness with insight and vision not known before in strict awareness.

The astonishment of the conscious in the presence of what is suddenly divulged by the unconscious is testified to by the concept of genius. Reverence toward genius is especially marked in modern romanticism but the word itself, and the general idea, are very old. A genius, of course, is originally a personal or familial spirit, worshiped along with the lares and penates, tutelary spirits of the household. The ancient attribution of insight to spirits and the modern attribution of insight to the unconscious testify in this respect to the same thing: a source beyond the limited, conscious ego. Reverence to one's own genius (however it is interpreted) is not unbecoming because it has power we do not consciously own, and it may leave us if treated improperly. There are too many instances of a source drying up, a barrier reared against its flow, leaving artists only with the technique and device, now sterile, which once allowed them to use inspiration to make art.

Inspiration may come to the artist with the force of revelation; literally he sees something he did not see before. And he may for the moment yield to it fully, merging his conscious self with his unconscious, experiencing almost mystic ecstasy. But the audience, too, has such moments: when an insight (not necessarily ideational) revealed by the object yields an especial response, or when there is resolution, or consummation, of what has gone before and the lines and movements come together in a perception of the whole; when recreation and attention have yielded discovery. The recurrent moments of utter concentration merge audience and object (and so in a way audience and artist) in an aesthetic experience of nearly mystical union in which the personality of the audience vanishes and only the object exists for consciousness. In a sense this is possession by the object: the periphery is gone and the object has expanded from the focus to possess the consciousness entirely.

Clearly, we are using the language of religion. But art is not religion, although the differences between the two are sometimes blurred. There are those who are religious for aesthetic reasons and those who are aesthetes for religious reasons, and there are permutations of the two. We discussed this in the last chapter, but here we should do it at greater length. Art can be talked about in the language of religion, as ideas can be talked about in the language of sensation, so differences, and mere metaphor, must be pointed out in some detail.

It is often not nearly so important to show that something is an error as to explain why people believe it. The reasons that art sometimes is a surrogate for religion and morals reveals something, of course, about man and society, but something also of the nature of art. Art does contain values and attitudes toward the world. It teaches, and its lessons endure. But it does not teach *the* truth; it teaches possible evaluations, attitudes, and truths. And art does bring its audience into intimate contact with a person, contact perhaps more internal than can be found in any other human relation. But it is a *human* relation and the contact is with a man, not God. No other product of human genius embodies the man as art does: Newton's *Principia* is surely on the highest level of human creation; but it does not embody the man as the Mass in B Minor does, or the ceiling of the Sistine Chapel. Gravitation would have to be dramatized in order to contain Newton's vision rather than his thought. It is never expression of thought that is intimately the man; rather it is expression of vision, which is the world seen by a man and the way a man sees the world.

The distinction between thought and vision is like the distinction between the statement of a belief and its expression in conduct. What is expressed in conduct may or may not be a belief that the actor could state in words; it is a belief integrated into the personality, part of the personality itself. So fully can we, at times, share a work of art with its creator that we can be said in a way to enter his mind (not, of course, in the particular circumstances of its creation, but only in his final vision as expressed in the product of creation). In those moments of utter concentration and understanding which constitute the fullest experience of art, in which we are more aware of the object of art than we are of ourselves, we *become* the artist insofar as we share the same vision and experience.

Sharing the world of the artist as we share his work, being led by him through our experience of that work, in a sense entering his mind, is either an analogue of mysticism or an instance of it. The mystical experience just as experience, and apart from interpretation, is quite similar in all religions; it is like, but no doubt more intense than, similar experience in love and art. The great differences are in interpretation: there are many explanations of the Presence we confront in mystical experience and of the preservation or dissolution of our own individuality. Most of this interpretation is based on prior theological belief and is just as likely to fit the mystical experience to those Procrustean patterns as it is to analyze the experience itself. So whether it is God we think we experience or Bach, some aspect of ourselves or the universe, is for the most part a matter of what we believe in our nonmystical moments.

Sometimes it is a person we feel through the work: it is Bach we love, not just the two-part inventions or the *St. Matthew Passion,* but the whole personal pattern of music we call Bach and mean when we use his name. It is, of course, Bach the artist we feel, and that is Bach the man only in a special sense. It is at once more intimately the man, and not the man at all, but what came from him and ceased to be in him. We can respond as we do because the artist's relation to his work is so intimate as to be an analogue of (a religious term again) Incarnation; the artist does not only make the object; in a sense he is in it and is revealed by it. It is not his body which is transformed into something else, but his person, in the most general sense of the word, that is extended by his work and expressed in it.

Another kind of aesthetic experience to which we can appropriately give a religious name, revelation, emerges as we are led through the aesthetic experience, in our encounter with meanings never met before, or never met in this way, meanings which strike us with the force of inevitability, because of their fitness in the world or context of the artist. A Mozart variation, for example, although one of many possible variations, makes us feel that this is the right one, appropriate to the world of Mozart as it might not be to the organization of emotions and meanings in the music of Schubert; or as a particular liquefaction or flow of color is almost necessary in the world of El Greco but would be out of place in the world of Dürer. The virtual inevitability of great art, the way in which each detail almost has to be what it is in the artist's context, brings immediate conversion, or acceptance of the artist's meaning, as self-evident propositions are accepted as soon as they are understood; there is no need for the persuasion, the argument, the evidence, which accompanies empirical statement. Rhetoric persuades; art at its best convinces, and convinces immediately. The temporary acceptance of the artist's world, the "willing suspension of disbelief," makes us ready to accept the detail in it. These worlds and their meanings are not ours, in that we have a feeling neither of discovering nor of inventing them, but of having them given us, of having been led to them by a power greater than, or different from, our own. The source of revelation, like the Presence in the mystical experience, is a matter of interpretation based on theories which precede the individual experience. But the feeling of revelation, of something revealed, is one of the responses to art as it is to religion.

The language of art can be sheerly technical: we can talk of strophe, diminished seventh, dénouement. And the technician may be rightly scornful of the orchestra conductor who tells his strings to play more like moonlight instead of discussing their fingering. But such language reveals little of aesthetic quality; no more than the technical language of engineering reveals of the beauty of bridges. The language of religion is not especially helpful when we deal with the *object* of art; it does illuminate the *experience* of it. A whole philosophy of civilization is implied when we try to explain the value of religious terms in the discussion of aesthetic experience, and this is not the place for such an undertaking. But it is worth pointing to some major alternatives. The language of religion may be thought of as sheer metaphor when applied to the audience of art, and its usefulness may be attributed to similarities in aesthetic and religious experience. Conversely, the same terms may be treated as deriving from the experience of art, and finding a place in religion because of the element of art in religion. A different tack altogether may be taken if we find in much of human experience types of relation and response which are appropriately labeled "revelation," "mysticism," "incarnation," etc. (as in insight, love, craftsmanship); and we may then argue that religion and art are areas of experience in which these types of relation and response are especially concentrated and systematized. Whichever direction we take, it is clear that such terms refer to recognizable experience which is thinned and distorted by thinking of it only in the language of behaviorist psychology.

Art as it exists in our day is independent of religion. Early secular art used the *form* of the religious rite to express either nonreligious interests or interests not acceptable as part of traditional religion. Elizabethan drama, for instance, evolved from an expansion of the ritual of the Mass. As the form first developed, it was used outside the church to dramatize Biblical stories, and later, when an elaborate and powerful theatrical form had been created, it was used for nonreligious material.

The pre-Elizabethan theater in England was dominated by the craft guilds, the butchers, grocers, shoemakers, and so on. No author's name was announced, the guild taking credit for every play it produced.[7] It is very likely that the plays were the result of collaboration. No matter how much any one man contributed, the age was not interested in individuals as artists and each play was thought of as a collective work. Bernard Berenson, art critic and historian, said, ". . . the notion of originality that has meant so much, so increasingly in the last two hundred years, scarcely existed prior to the first dawn-

[7] A typical announcement of a play performed by a guild was: *The Creation of Eve, with the Expelling of Adam and Eve out of Paradise, Acted by the Grocers of Norwich.*

ing of the so-called Romantic period. Indeed, it would be interesting to know whether as a category, and as a quality, it was recognized at all. . . ."

Generally, it is common for art when it is first separated from ritual to be collective, at least in the sense that individual names are not associated with it. The churches of the Middle Ages were collective efforts, as were folk songs and folk epics—the Norse sagas, the Mabinogion, the Nibelungenlied, the Song of Roland. Collective art is close to ritual—if not the ritual of institutional religion, then the ritual of communal traditions and folkways.

At a further remove from ritual is individual art, each work being made by one man and recognized as such. The contrast between ritual art and individual art is a sharp one. Once a ritual form is set it remains as definitively the same as succeeding generations can keep it. Even the quality of performance in each presentation of the ritual is as little varied as possible. New interpretations are not permitted and the training of performers is relatively unchanging. After all, ritual is not just a presentation or dramatic performance. It is sanctioned by religion and is usually thought to have some efficacy to ensure a good harvest, or success in battle, or some other sign of the gods' favor.

One kind of individual art, probably the earliest in time, is the occasional: art produced for a specific place or occasion—the death of a king, a triumphal procession, a state holiday. Court music was usually occasional, and so was classical Greek sculpture, which was made for a particular place: a public square, a temple, the scene of a battle.

Occasional art celebrated or commemorated *this* event at *this* time. Ritual was more general; it celebrated a *kind* of event—birth, not *a* birth; marriage, not *a* marriage. Ritual was repeated more or less identically at specified intervals. Occasional art might be repeated, as in the case of a play, but that was because of its merit or popularity; it was made for a single occasion. Yet occasional art was a substitute for religious ritual or an extension of it. It dramatized or celebrated a secular event not incorporated in religion or a particular event which was celebrated in ritual only as a type and so did not include reference to an actual person or date. The drama of ancient Greece was occasional art which contained the ritual forms but developed them in original ways. It was performed at religious festivals which included the entire community, but new plays were written every year; Greek drama was like a ritual created afresh annually by individual artists.

The dramatic poet was highly individuated, but he had a function in the community which was for the sake of the community. Even if he was critical of his own land, like the tragic poet Euripides or the comic poet Aristophanes, it was to instruct the community, of which he felt himself fully a member.

Individuality easily passes into eccentricity when it is severed from the social ties which give it relevance to the life of its time. The artist of our own day is no more individual than Euripides or Aristophanes, but he is less sustained by common bonds and roots; often, indeed, he is alienated. Daily book reviewers and writers of articles for Sunday supplements make easy scandal of the distance between what they think of as the common man and the artist. When the United States Information Service chooses American books for its libraries abroad, questions are raised about the image of America put before foreign eyes. Faulkner's Snopeses and Bundrens, Steinbeck's Okies, Nabokov's Humbert Humbert are not God's noblest creatures and the land they inhabit no longer seems the last best hope of earth. It is a far cry from Aeschylus' radiant hopes for Athens and Sophocles' passion for her. And how strange in contrast are Shakespeare's words:

*This happy breed of men, this little world,*
*This precious stone set in the silver sea,*
*Which serves it in the office of a wall*
*Or as a moat defensive to a house*
*Against the envy of less happier lands—*
*This blessed plot, this earth, this realm,*
*this England.*

But the alienated man always sees his country as if from the outside; he is in it, but not of it. And he is, almost by nature, critical. Whatever blame—if any is to exist—that can be attributed to the artist is slight compared with the blame attaching to what should be his public. Yet what has happened is the result of a social upheaval, so blame is irrelevant. The audience for serious art is very small, it is distributed over a large country, it is not cohesive, and (except for art collectors, who are often mere investors) it is not extremely wealthy. Surrounding this audience is an ocean of the merely literate, the philistine, and the apathetic, products not of schooling but of compulsory education.

Universal literacy combined with money and leisure have produced a bottomless market for entertainment and, in consequence, entertainment is the largest industry in the United States today. The difference in the market for, say, literary art and for literary entertainment can be seen in sales figures like these: a truly distinguished American poet in his middle years may have all his verse published in one volume as Collected Poems, and sell 3,000 copies; a best-selling popular novel may sell half a million copies. And even in our colleges, where the study of English is required, and there are more students in English classes than in any other single study, it is the literature of the past that is read, almost to the exclusion of the literature of the present.

The more thorough the alienation of the artist, the more likely that artists will make a sub-culture of their own. And this can only widen the breach between artist and public, making the artist more critical and the public more resentful. But even the alienation of the artist, hard as it is for him to bear, serves a function. Instead of expressing the major values of his culture, in the manner of so many of his forebears, the contemporary artist becomes the conscience of his time, crying havoc, and illuminating the gap between avowed ideals and sordid practice. Behind the figure of Flem Snopes is the towering shadow of Thomas Jefferson. T. S. Eliot writes these lines:

*The person in the Spanish cape*
*Tries to sit on Sweeney's knees*

*Slips and pulls the table cloth*
*Overturns a coffee cup,*
*Reorganized upon the floor*
*She yawns and draws a stocking up*

and ends the same poem:

*The nightingales are singing near*
*The Convent of the Sacred Heart,*

*And sang within the bloody wood*
*When Agamemnon cried aloud,*
*And let their liquid siftings fall*
*To stain the stiff dishonoured shroud.*

Demands made on the serious artist today often overlook the industrial revolution, the mushrooming of population, mass education, and all else that has come between him and Dante. He is expected to be seer, prophet, and teacher, not just another human who worries about paying the rent. And he is mistakenly expected to amuse or delight, which he is often capable of doing in part, but not entirely. For, unlike the work of writers, painters, and composers who are in the entertainment industry, the work of the artist today, as always, is difficult and requires serious attention. The entertainer does not see a world uniquely and afresh, but concocts a world out of the stereotypes of his time. His audience understands him even when it is half asleep. But the perspectives and values of the artist are sufficiently his own that his audience must struggle to grasp a novel world. And the concerns of the artist are fundamental to life, not mere play with its surface, so the audience is faced with life, which it must see as its own. It is not a light and pleasant prospect to sit alone in a room with *King Lear;* it is terrifying, however wonderful.

The additional difficulties, beyond those of art itself, of some of today's alienated artists are real enough, but they are probably less the cause of the small size of the serious audience than the conditioning of the public. For the public is so used to the banal and the stereotyped as it flickers on theatre and home screens, comes across the footlights, screams from the jukeboxes, and is ground out by the presses, that it is not willing to give to art the discipline and effort that it reserves for work and money.

The relation between the artist as artist and the artist as man has been a puzzle for centuries, and attempted explanations have resulted in theories of widely different kinds. An ancient theory virtually identified the poet and the prophet, contrasting them sharply with the philosopher and the practical man. The artist was assumed to have divine powers which not only gave him especial insight but enabled him to foretell the future. He was something of a holy man and was expected to live as such. Some ages have distinguished sharply among the arts; in classical Greece, for example, the poet was an artist while the sculptor was an artisan.

Style, we are often told, is the man.[8] Of course it is—and of course it is not. The man he is in everyday life is rarely in the artist's style

[8] "The style is the man himself," said the Comte de Buffon in his *Discourse on Style.*

at all, and then only slightly. The artist as artist is a different man from the artist as citizen, husband, or father. There is no other way of accounting for the gap between the man we know in other activities and the man we know as artist. Poor, distraught Vincent Van Gogh of the self-amputated ear is far from the triumphant painter of the last canvases. The test is simple. Examine a work of art by a man of whom you know nothing, see what you can predict about his person, then compare the predictions with the accepted biographical data. The disparity will be enormous, or, put differently, knowledge of the work will prove irrelevant to knowledge of the life.[9]

The proof of the relevance of biographical work would come, conversely, if we could predict a man's work from the study of his life. We can say things about the relation of Keats's poems to the status of his affair with Fanny Brawne, but can we predict from the latter alone anything about poems we have not already read? Can we even predict that Keats would write poetry if, as a matter of fact, we did not know it? What has the battle of Lepanto to do with *Don Quixote*? Knowing the book's existence anyone can guess; Chesterton guessed that Don John of Austria was the model for the sorrowful knight. But there is no reason, on the basis of Cervantes' part in the battle, to predict that there would be Quixote at all.[10]

Yet we feel that we know a man intimately when we know his work, recognize his brush stroke, identify a new work as his by his quality, his style—copyable perhaps, but uniquely his. We do know and identify him as artist, by his use of the forms, the conventions, the logic of his art, by his techniques and devices. And we know him in other ways. We feel the emotional tone, share the vision, identify the quirks and crotchets of attitude and mind. Surely we are right in feeling that we know him because our knowledge is by way of a medium more expressive, more revealing, than conduct usually is.

Art calls forth, in men who can be artists, facets of the inner life that are regularly suppressed and diverted in daily living. Whatever we call the wellsprings of life and thought, the unconscious, perhaps, they enter consciousness and are controlled and directed in art, not re-

[9] Perhaps much of the controversy about the authorship of the works printed in the name of Shakespeare is due to a belief that the author must, in his daily life, have exhibited the genius and temper of the plays.

[10] An important qualification of these arguments about life and work is that when they are both known, either may be *useful* in understanding the other.

pressed. So we can have deeper insight into the man through his art than through personal intimacy, know him better in a variety of ways than his own family does, if they know him only as man, not as artist.

There are men who find, in some few activities of life, the kind of demand on personality that art makes, combined with the kind of freedom that it allows. For these men the resources of the self may be utilized in action somewhat as they are utilized by the artist in his work. And their personalities may expand, develop, be altered, by the demands and possibilities of life. So, too, the personality of the artist is made during the creative process. The artist as artist is changed, not only in a lifetime of creation, but from the beginning to the end of a single work, just as the work itself is changed in all its stages. The artist, to be sure, makes the object, but the making of the object develops the artist. What artist ever began with an exact blueprint of a work in mind, and then carried it out with no change or growth? Such may be a persuasive theological description of the relation of God's mind to his work, but it is quite inadequate to the facts of man's creative activities.

There is an old theological question of great difficulty: If God is perfect, why did he create man? Creation seems to imply a lack of perfection, a need for something. The same question can be applied to art: Why does the artist create? Does he need something he does not have in his life? Art, then, might be a compensation in imagination for what is lacking in daily existence. And the artist might be a man whose life was insufficient, or who was aware of insufficiency and capable of making up for it. In one popular theory, based on Freud, the artist is a neurotic who keeps himself healthy so long as he creates art, which is his therapy. Lionel Trilling has called attention to the advantage of "the myth of the sick artist." "To the artist himself," he says, "the myth gives some of the ancient powers and privileges of the idiot and the fool, half-prophetic creatures, or of the mutilated priest." To the philistine, the myth permits not listening to the artist, or listening without hearing. And those of a third group, sensitive and sympathetic to art, and somewhat neurotic, sanction their own situation by accepting the myth and by living as they imagine artists do, even making vague gestures in the direction of artistic work.

There seems no good reason to think of artists as more neurotic than other creative people—scientists, for example. And there is a real question whether they are more neurotic than butchers, bakers, or garage mechanics. Apart from the advantages of the myth of the sick artist, though, there is a good reason why the artist may be thought more neurotic than he is. Creativity is an awesome phenomenon when we observe it in anyone—philosopher, mathematician, scientist. But it is touched with something more when we see it at work in the artist. For he alone among men creates a world of his own, and thus exhibits in imagination and in a few artifacts, a power like that we attribute really to divinity. Those who have such power may have to be denigrated that they be not too much envied.

# Symbols & Civilization

The world is a flux of more or less identifiable things and events in which a solitary animal may, with luck, pick his way and preserve his life. But the "world" in which we live is quite different. It is made up of separate and identifiable components, it is rich in meanings, and much of it can be manipulated for our purposes. It is not just a world into which we were born, but a world we inherited. Its parts have been characterized and distinguished from each other, its relations noted and interpreted, its meanings expressed and analyzed, by our forebears. Thus society has, to some extent, created the world we perceive, its shapes, its meanings, its beauties and uglinesses. We inherit, by being born in society, an intelligible world and a system of symbols, and we cannot have one without the other.

In its earliest and most basic uses, language seems to have made the world intelligible by relating it to human activities and purposes. Although we commonly assume that we define words by examining the things they stand for and noting similarities among them, this is only part of a process which includes selection of those similarities we want to emphasize. And it is only one of several ways of defining, perhaps a recent way. When an Indian tribe whose livelihood was based on agriculture used the same word for dancing and working, they were probably using the word to mean "the way to get a good crop," for the religious dance seemed at least as important for that purpose as work in the fields. Malinowski, writing of the Trobriand Islanders, says, "A word *means* to a native the proper use of the thing for which it stands, exactly as an implement *means* something when it can be handled and means nothing when no active experience is at hand."

In the tradition of Plato and Aristotle it was believed that we could and should find the single real meaning of things, their "essences." Today we are aware that there are no such "real" or "single" meanings, and we deliberately ascribe meaning to objects in terms of human use. But our uses are more numerous and varied than those of primitive man. One of our new uses is science, and in science the meaning we select is one of many possible meanings; it is chosen because of what we want to do with it *intellectually*. "Functional" meanings and "operational" meanings clearly depend on our behavior in social usage and scientific testing. A chair is an object designed for man to sit upright in, and that is the definition of "chair." But even the most seemingly objective "structural" meanings are the result of selection of those structural features which suit our systems of taxonomy or our particular concerns. A whale is a "mammal," not a "fish," because zoology classifies in terms of vertebrae and mammary glands rather than natural habitat.

A symbol is a surrogate for a thing or event. It wrests a thing from sensuous immediacy and fixes it in the mind, makes it an "object," to be thought of at any time. We understand things only when we place them as instances of a type, or cases of a law. Symbols can be so placed and classified by the mind, and they can be organized in processes of generalization and deduction. The things for which the symbols stand are not so tractable and are understood only through their surrogates.

In daily activity, when *A* talks to *B* about something in experience, he calls attention to one of its innumerable aspects. This ensures that *B* will see it as *A* does, or will see the same aspect. Not only has perception been standardized by a common symbolic heritage, it focuses now on one thing, now on another, as communication directs it. The social presupposition of communication is the existence of a group which has identified specified meanings with particular symbols. Because the identification is common, men can co-operate for common ends, can transmit a culture, and can enhance the experience of things by adding meaning to them and sharing meaningful experience.

Symbols serve social purposes when they are defined in terms of personal and group usage of the objects they stand for and of the relations of objects to us. They serve the purposes of science when they are defined in terms of intellectual use and the relations of objects to each other. They serve the purposes of art when they are used in all these meanings plus the relation of the objects to imagination.

Magic implies the belief that symbols are a part of the things they stand for, or a ghostly essence, a kind of soul, of those things. Things can be controlled by the words that stand for them, by images of them, even by an imprint, like the impression a body leaves on a bed. For magic, manipulation of symbols is a manipulation of things. People come when we call their names. Why should not spirits or material objects? Magic is inimical to science, which must distinguish between the intellectual manipulation of thought and the physical manipulation of experiment. And it is inimical to art, which must distinguish the play of imagination from the movement of things. But magic, at least in a metaphoric sense, is still vital to art, for the symbols of art are more than signs and are properly treated as material objects in themselves. The scientist uses the most convenient word, but the poet uses the "right" one and is concerned with its sound and cadence. The symbols of science are just symbols, but the symbols of art are also materials of which a work is built, objects to be apprehended and enjoyed.

The interaction of society with the symbolic worlds of science and art is continuous. Particular social conditions make science possible, and science in turn alters society by the knowledge it attains and the technology it yields. Art exists under all social conditions, but the kind and quality of art depend on the conditions of society. In turn, art alters the perceptions and standards of society. Education, which transmits the culture, transmits a new world to each generation when science and art are active. The circumstances of human life change, to be sure, under the impress of economic and political change. The "material" conditions of life have seemed to many men to provide the base for all variation in "spiritual" and "intellectual" conditions. But the influence is neither so simple nor so limited in its direction. Economics and politics are activities of men who are necessarily molded by the symbolic world of meaning, knowledge, and value that they inhabit. And they choose among alternatives not just in accordance with the push of material forces, but also in accordance with their morals and values, their stereotypes and ideals, their understanding of the situation that confronts them, and the consequences of their actions.

# Selected Bibliography


## SCIENCE

Norman Campbell: *What Is Science?*
Philipp G. Frank, ed.: *The Validation of Scientific Theories*
Walter B. Cannon: *The Way of an Investigator*
Morris R. Cohen and Ernest Nagel: *An Introduction to Logic and Scientific Method*
Ernest Nagel: *The Structure of Science*
Bernard Barber: *Science and the Social Order*
R. B. Braithwaite: *Scientific Explanation*
Alfred Tarski: *Introduction to Logic*
Herbert Feigl and May Brodbeck, eds.: *Readings in the Philosophy of Science*
George Sarton: *A History of Science*

## HISTORY

R. G. Collingwood: *The Idea of History*
Sidney Hook: *The Hero in History*
F. J. E. Woodbridge: *The Purpose of History*
Benedetto Croce: *History as the Story of Liberty*
C. V. Langlois and Charles Seignobos: *Introduction to the Study of History*
José Ortega y Gasset: *Toward a Philosophy of History*

## ETHICS

John Dewey: *Human Nature and Conduct*

C. D. Broad: *Five Types of Ethical Theory*
Charles L. Stevenson: *Ethics and Language*
G. E. Moore: *Principia Ethica*
F. H. Bradley: *Ethical Studies*

## LANGUAGE AND MAN

Otto Jespersen: *Growth and Structure of the English Language*
Benjamin Lee Whorf and J. B. Carroll: *Language, Thought, and Reality*
Edward Sapir: *Language*
Edward Sapir: *Culture, Language, and Personality*
Ernst Cassirer: *Language and Myth*
Ernst Cassirer: *An Essay on Man*

## RELIGION

Gilbert Murray: *Five Stages of Greek Religion*
W. K. C. Guthrie: *The Greeks and Their Gods*
William Robertson Smith: *The Religion of the Semites*
Fustel de Coulanges: *The Ancient City*
Emile Durkheim: *Elementary Forms of the Religious Life*
Max Weber: *The Protestant Ethic and the Spirit of Capitalism*
Sigmund Freud: *Totem and Taboo*
Sigmund Freud: *The Future of an Illusion*
William J. Goode: *Religion Among the Primitives*

## ART

Benedetto Croce: *Aesthetic*
John Dewey: *Art as Experience*
George Santayana: *The Sense of Beauty*
George Santayana: *Reason in Art*
R. G. Collingwood: *The Principles of Art*
Bernard Berenson: *Aesthetics and History*
Erwin Panofsky: *Meaning in the Visual Arts*

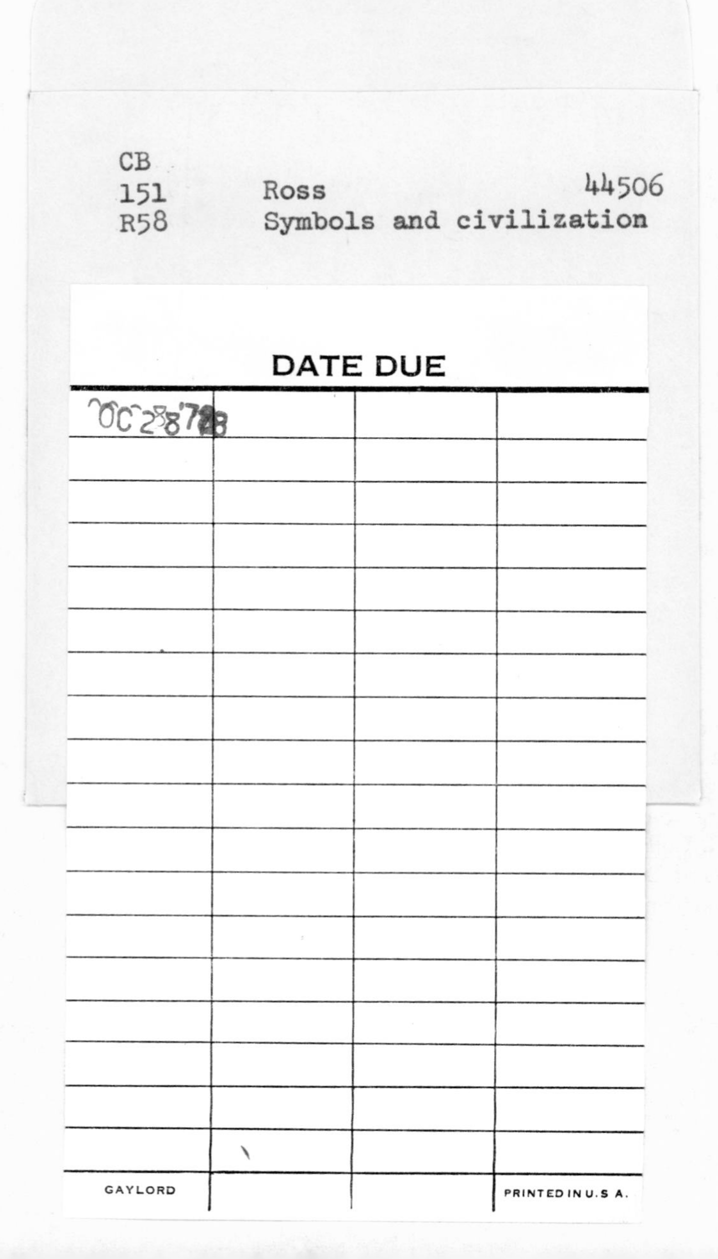

CB
151
R58
Ross
44506
Symbols and civilization
DATE DUE
GAYLORD
PRINTED IN U.S.A.